Clime Chronicles: Navigating the Pathways of Climatology

Welcome to a journey that transcends time and geography, a voyage that uncovers the hidden stories woven into the fabric of our planet's climate. In the pages of "Clime Chronicles: Navigating the Pathways of Climatology," we embark on an expedition through the intricate tapestry of Earth's climatic systems, exploring the past, present, and potential future of our global climate.

Climatology is more than just the study of weather patterns; it's an intricate puzzle that connects oceans, atmosphere, land, and life in a delicate dance that shapes the conditions we experience daily. It's a discipline that delves into the mysteries of climate change, deciphers the forces behind extreme events, and offers insights into the very essence of Earth's history.

As we embark on this journey, we will traverse diverse landscapes – from the icy realms of polar climates to the lush embrace of tropical ecosystems. We will uncover the whispers of ancient ice cores, interpret the language of temperature records, and unlock the secrets held within historical climate data. We'll learn how humanity's actions have shaped and reshaped the climatic narrative, and we'll explore the pathways to a more sustainable future.

Our expedition is not just a scientific exploration; it's a call to action. It's an invitation to understand the delicate balance that sustains life on our planet and to appreciate the intricate interplay of factors that influence our weather, seasons, and environment. It's an acknowledgment of our shared responsibility to preserve the planet's climatic harmony for generations to come.

As we turn the pages of "Clime Chronicles," let us become both travelers and stewards. Let us unravel the threads of climate science, engage in meaningful discussions, and step into a world where knowledge empowers us to make informed choices. The voyage ahead is filled with insights, challenges, and discoveries that will broaden our horizons and deepen our connection to the vast symphony of Earth's climatic symphony. Welcome to the captivating world of climatology – where each chapter unveils a new facet of our planet's climatic narrative.

Introduction

- The significance of climatology in understanding Earth's climate systems.
- An invitation to explore the diverse aspects of climatic patterns and change.

Chapter 1: Foundations of Climatology
- Defining climatology and its relationship with meteorology and climate science.
- Historical development and evolution of climatology as a scientific discipline.
- Importance of climate data collection, analysis, and interpretation.

Chapter 2: Earth's Climatic Zones
- Exploration of different climatic zones, including polar, temperate, tropical, and arid regions.
- Factors influencing variations in temperature, precipitation, and atmospheric conditions.
- Ecosystems and biodiversity unique to each climatic zone.

Chapter 3: The Dance of Climate Elements

- Atmospheric circulation patterns and their impact on weather and climate.
- Ocean currents, their role in heat distribution, and their influence on regional climates.
- Interactions between land, sea, and atmosphere that shape climate characteristics.

Chapter 4: Historical Climate Records

- Uncovering historical climate data sources, such as ice cores, tree rings, and sediment layers.
- Using proxies to reconstruct past climates and understand long-term climate trends.
- Insights into natural climate variability and patterns over centuries and millennia.

Chapter 5: Climate Change and Global Warming

- The concept of climate change and its driving factors, including greenhouse gas emissions.
- The role of human activities in accelerating global warming and altering the climate.
- Impacts of climate change on ecosystems, sea levels, and extreme weather events.

Chapter 6: Paleoclimatology: Unraveling Ancient Climates

- Exploring paleoclimatology as a window into Earth's climate history.
- The study of past climate conditions through geological records and proxy data.
- Insights into ancient climates, ice ages, and shifts in atmospheric composition.

Chapter 7: Adapting to a Changing Climate

- Strategies for adapting to changing climatic conditions and mitigating risks.

- Case studies of communities, industries, and regions implementing climate adaptation measures.
- The importance of resilient infrastructure, sustainable agriculture, and disaster preparedness.

Chapter 8: Climate Modeling and Prediction
- The role of climate models in simulating past, present, and future climate scenarios.
- Techniques for predicting climate patterns, trends, and potential impacts.
- Challenges and uncertainties in climate modeling and the importance of improving accuracy.

Chapter 9: The Human Connection to Climate
- Exploration of the historical and cultural significance of climate in different societies.
- Societal impacts of climate change, including migration, conflicts, and economic shifts.
- The role of education and awareness in fostering responsible climate stewardship.

Chapter 10: From Knowledge to Action
- The imperative for global cooperation in addressing climate change.
- Policies, agreements, and initiatives aimed at reducing greenhouse gas emissions.
- Individual and collective actions to mitigate climate change's effects and ensure a sustainable future.

Chapter 11: Emerging Trends and Innovations
- Advances in climate science and technology, including remote sensing and data analytics.
- Innovative solutions for sustainable energy, carbon capture, and climate adaptation.
- The potential for science and innovation to drive positive change and shape policy.

Conclusion: A Call to Climate Stewardship

- Reflecting on the insights gained from the exploration of climatology and paleoclimatology.
- Emphasizing the importance of responsible climate stewardship for future generations.
- Encouraging readers to engage in ongoing learning, awareness, and action to safeguard Earth's climate.

Epilogue: Continuing the Journey

- Acknowledging the ever-evolving nature of climatology.
- Inspiring readers to carry the knowledge and lessons of climatology into their lives and communities.
- A call to continue exploring, learning, and advocating for a resilient and sustainable planet.

The significance of climatology in understanding Earth's climate systems

The Significance of Climatology: Illuminating Earth's Complex Climate Systems

In the intricate tapestry of Earth's natural processes, climatology emerges as a guiding thread, offering profound insights into the dynamics that govern our planet's climate systems. Climatology is more than a study of temperature and precipitation; it's a multidisciplinary science that unlocks the secrets of our past, present, and potential future climates. Its significance is profound, as it illuminates the delicate balance that sustains life, shapes ecosystems, and influences the course of human history.

Unveiling Earth's Climate Patterns: Climatology unveils the intricate patterns that govern our planet's climate. From the rhythmic dance of ocean currents to the majestic sweep of atmospheric circulation, climatology unravels the forces that define temperature, humidity, wind, and other crucial variables. This understanding allows us to predict weather patterns, anticipate climate anomalies, and foster resilience in the face of environmental challenges.

Historical Context and Paleoclimatology: The past whispers through the pages of climatology, as it uncovers historical climate records and deciphers the stories embedded in ancient ice cores, tree rings, and sediment layers. By examining these archives, climatologists piece together Earth's climatic history, revealing epochs of warmth, ice ages, and dramatic shifts. This historical context offers a sobering reminder of the dynamic nature of our planet's climate.

Climate Change Understanding: In an era of rapid environmental change, climatology is a cornerstone in understanding climate change dynamics. It dissects the factors that contribute to global warming, examines the role of greenhouse gases, and identifies the intricate web of interactions that drive temperature variations and sea-level rise. Armed with this knowledge, we can evaluate the impacts of human activity and make informed decisions to mitigate their effects.

Safeguarding Ecosystems and Biodiversity: Ecosystems, from polar tundras to tropical rainforests, are intricately tied to climatic conditions. Climatology helps us comprehend the relationship between climate and biodiversity, revealing how shifts in temperature and precipitation patterns can disrupt delicate ecosystems. By understanding these connections, we can develop strategies to conserve biodiversity and safeguard fragile habitats.

Predicting Extreme Events: Extreme weather events, from hurricanes to droughts, hold the power to reshape landscapes and communities. Climatology enhances our ability to predict and prepare for these events by identifying the atmospheric conditions that foster their formation. This knowledge empowers us to develop early warning systems, implement disaster preparedness plans, and minimize the impact of such events.

Shaping Policy and Sustainability: Climatology empowers informed decision-making by governments, industries, and communities alike. It underpins climate policies, shapes sustainable resource management, and guides initiatives for reducing carbon emissions. Armed with climatological insights, societies can adapt to changing conditions, foster environmental stewardship, and strive for a more sustainable future.

In essence, climatology isn't just a scientific endeavor; it's a bridge between understanding and action. It beckons us to unravel the intricate threads that weave our planet's climate fabric and empowers us to shape a future that embraces sustainability and resilience. As we delve into the mysteries of climatology, we unlock the keys to Earth's climatic complexities – keys that hold the potential to safeguard our planet and all the life it sustains.

An invitation to explore the diverse aspects of climatic patterns and change

Embarking on a Journey: Explore the Diverse Aspects of Climatic Patterns and Change

Step into a world of wonder, where the skies above and the earth beneath are engaged in an eternal dance – a dance that shapes the intricate patterns of climate that envelop our planet. The invitation is extended to you, an explorer of the curious, a seeker of knowledge, to embark on a voyage through the captivating realm of climatic patterns and change.

Unveiling the Mysteries: Behind every gust of wind, every drop of rain, and every sunlit day lies a realm of mysteries waiting to be unraveled. The symphony of climatic patterns and change is a tapestry woven from the interactions of land, oceans, atmosphere, and life itself. Your journey begins by pulling back the veil on this symphony, uncovering the forces that orchestrate Earth's climate ballet.

A Tapestry of Diversity: Climatic patterns span the globe, each region painted with its own colors, rhythms, and stories. From the icy realms of the poles to the lush canopies of tropical rainforests, climatic diversity is a testament to the planet's rich palette. As you traverse diverse landscapes and ecosystems, you'll witness the delicate balance that sustains life and marvel at the interconnectedness of it all.

Time's Echoes: History whispers through the ages, carried on the winds of climate change. The echoes of past climates resonate in ancient ice, ancient forests, and sediment layers. These whispers speak of epochs long gone, of ice ages and interglacial periods, offering a glimpse into Earth's ever-changing climatic narrative. Let your journey through climatic patterns echo with the wisdom of ages past.

Faces of Change: Change is a constant companion in the world of climatology. It paints its mark through shifts in temperature, ocean currents, and atmospheric compositions. As you delve deeper, you'll uncover the intricate relationship between human activities and the changing climate. You'll witness the profound impacts of global warming, prompting contemplation about the actions needed to shape a sustainable future.

Seeking Solutions: With knowledge comes empowerment, and with empowerment comes the responsibility to act. Your journey through climatic patterns and change extends beyond exploration; it's a call to action. It's a call to champion climate stewardship, to advocate for sustainable practices, and to stand as a guardian of Earth's delicate climatic equilibrium.

Connecting to the Web: As you journey through the diverse tapestry of climatic patterns, you'll discover your place in the web of life. You'll see how every action, every choice, reverberates through this interconnected ecosystem. Your exploration is an affirmation of our shared responsibility to preserve and protect this delicate balance for present and future generations.

A Call to Explore: The invitation is clear – explore the myriad facets of climatic patterns and change. Embrace the beauty of Earth's diverse climates, delve into the history that shapes our present, and recognize the power of human agency in shaping our future. Let your exploration be a catalyst for understanding, action, and the stewardship of a world that depends on the delicate symphony of climatic harmony.

Defining climatology and its relationship with meteorology and climate science

At the heart of our planet's intricate climatic tapestry lies the science of climatology – a discipline that unravels the threads of Earth's long-term climate patterns, revealing the profound interplay between atmospheric elements, ocean currents, and the dynamics of life. While often intertwined, climatology, meteorology, and climate science are distinct yet harmonious fields that collectively illuminate the story of our planet's climate.

Climatology: Unveiling Earth's Climate Narrative Climatology is the scientific exploration of Earth's climates – the patterns of temperature, precipitation, humidity, wind, and more that shape the conditions we experience over extended periods. It delves into the deeper currents of climate systems, deciphering the forces that drive changes and fluctuations over decades, centuries, and even millennia. By examining past, present, and future climates, climatology offers a holistic perspective on the planet's climatic rhythms and their implications.

Meteorology: Deciphering the Dynamics of Weather Meteorology, the sibling of climatology, centers on the immediate and short-term atmospheric conditions that give rise to weather. It investigates the processes that form clouds, generate rainfall, and create the daily conditions we observe. While climatology takes a broader, long-term view, meteorology examines the intricate dance of atmospheric elements that result in the weather we experience on a day-to-day basis.

Climate Science: Bridging Past, Present, and Future Climate science is the nexus where climatology and meteorology converge, creating a bridge between historical climate data and immediate weather phenomena. It studies the patterns and trends in temperature, precipitation, sea levels, and atmospheric composition to comprehend the broader shifts and long-term impacts of our changing climate. Climate science connects the dots between the past, present, and projected future states of Earth's climate systems.

Synergy and Complementary Insights: The relationship between climatology, meteorology, and climate science is one of synergy, where each field contributes unique insights to the larger narrative of Earth's climate. Climatology, with its focus on long-term trends, enriches our understanding of the background against which meteorological events unfold. Meteorology, in turn, provides the dynamic context in which climatic changes manifest in our daily lives. Climate science brings these perspectives together, allowing us to perceive the intricate tapestry of Earth's climate in its entirety.

A Holistic Understanding: As we navigate the complexities of our planet's climate systems, it becomes clear that climatology, meteorology, and climate science are three integral facets of a holistic understanding. Together, they empower us to grasp the full spectrum of atmospheric phenomena – from the intricacies of short-term weather to the grand symphony of climatic changes spanning centuries. Through their combined efforts, we gain the tools to comprehend, adapt, and respond to the intricate rhythms that define our world's climate narrative.

Historical development and evolution of climatology as a scientific discipline

Tracing the Evolution of Climatology: From Ancient Curiosity to Modern Science

The story of climatology is a journey that spans millennia, driven by human curiosity, observation, and an insatiable desire to unravel the mysteries of Earth's climate. From the earliest civilizations to the present day, the development of climatology as a scientific discipline has been marked by profound discoveries, paradigm shifts, and the convergence of diverse fields.

Ancient Foundations: Curiosity and Observation Long before the formalization of climatology as a scientific field, ancient civilizations recognized the patterns of weather and climate. Early societies, such as the Egyptians, Greeks, and Chinese, documented climate variations and their impact on agriculture, trade, and daily life. These observations laid the foundation for understanding the dynamic relationship between the environment and human activities.

The Birth of Climatology: Enlightenment and Measurement The Renaissance era witnessed a transformation in the study of climate, as thinkers like Leonardo da Vinci and Galileo Galilei embraced empirical observation and measurement. The Age of Enlightenment brought about an intellectual awakening, and scientists began recording weather data systematically. Notably, the invention of the barometer by Evangelista Torricelli in the 17th century marked a significant step in quantifying atmospheric pressure and its role in climate dynamics.

19th Century Advancements: Exploration and Synthesis The 19th century witnessed the blossoming of climatology as an interdisciplinary field. Explorers and naturalists, including Alexander von Humboldt, embarked on expeditions to diverse regions, collecting data on temperature, altitude, and vegetation. Their efforts laid the groundwork for synthesizing climatic patterns on a global scale and understanding the factors that influence climate variation.

The Emergence of Modern Climatology: 20th Century The 20th century marked a turning point in climatology's evolution, with technological advancements and a growing recognition of human impact on the climate. The establishment of global networks of weather stations allowed for more comprehensive data collection and analysis. The recognition of the role of greenhouse gases in the greenhouse effect, pioneered by scientists like Svante Arrhenius, set the stage for understanding climate change.

Interdisciplinary Collaborations: From Meteorology to Climate Science As the 20th century progressed, climatology became increasingly intertwined with meteorology and other scientific disciplines. The emergence of climate models, supercomputing, and satellite technology enabled researchers to simulate and predict climate behavior. The establishment of the Intergovernmental Panel on Climate Change (IPCC) in the late 20th century marked a milestone in international collaboration to address climate change.

The 21st Century and Beyond: A Call to Action In the 21st century, climatology faces the challenge of addressing rapid climate change driven by human activities. The evolution of climatology continues to emphasize the interdisciplinary nature of the field, with collaborations between climatologists, meteorologists, geologists, ecologists, and policymakers. Advances in data analytics, remote sensing, and climate modeling continue to shape our understanding of complex climatic systems.

Conclusion: A Legacy of Discovery and Responsibility The historical development of climatology reflects humanity's profound connection to the environment and the quest to comprehend the intricate dance of Earth's climate. From ancient observations to modern scientific rigor, climatology's journey is a testament to human curiosity, innovation, and the recognition of our shared responsibility to safeguard the delicate balance of our planet's climate systems.

Importance of climate data collection, analysis, and interpretation

Harnessing the Power of Climate Data: Understanding the Vital Importance

In the realm of climatology, data serves as both the compass and the map, guiding us through the intricate landscape of Earth's climate systems. The collection, analysis, and interpretation of climate data are not mere scientific endeavors; they are essential pillars that underpin our understanding of the past, present, and future of our planet's climate. Their importance cannot be overstated, as they provide the foundation for informed decision-making, mitigation strategies, and a sustainable future.

Revealing Historical Insights: Climate data collection, spanning centuries and millennia, offers a portal to the past. Ancient ice cores, sediment layers, tree rings, and historical documents unveil a rich tapestry of climate variations. By deciphering these archives, scientists can reconstruct climatic conditions long before the era of modern instrumentation. This historical context provides insights into natural climate variability and lays the groundwork for understanding our planet's dynamic climate history.

Charting Contemporary Trends: In the present day, an array of weather stations, satellites, buoys, and sensors contribute to an extensive web of climate data collection. These instruments record temperature, precipitation, sea levels, atmospheric composition, and other key variables. The analysis of this contemporary data illuminates short-term weather patterns and long-term climate trends, allowing us to discern the impacts of human activities on the environment.

Predicting Future Trajectories: Climate data serve as crystal balls, granting us glimpses into potential future scenarios. Through climate modeling and simulation, researchers can project the consequences of various greenhouse gas emission scenarios, enabling us to anticipate changes in temperature, sea levels, and extreme weather events. This predictive power empowers us to formulate strategies for adapting to and mitigating the effects of climate change.

Driving Informed Decision-Making: Governments, industries, and communities rely on climate data to make informed decisions. Agricultural practices, urban planning, disaster preparedness, and resource management all hinge on a deep understanding of local and global climate patterns. With accurate and comprehensive data, policymakers can develop strategies that enhance resilience and sustainability in the face of changing climatic conditions.

Advancing Scientific Understanding: The analysis of climate data fuels scientific discovery, fostering collaborations across disciplines. Climatologists, meteorologists, geologists, oceanographers, ecologists, and more, work together to decipher the intricate puzzle of Earth's climate. Shared data pools lead to innovative research, enabling breakthroughs in understanding the complex interactions between atmospheric, oceanic, and terrestrial systems.

Fostering Public Awareness: Climate data isn't confined to laboratories; it is a vital tool for public awareness and education. Data visualizations and accessible summaries help communicate the urgency of climate change to diverse audiences. Armed with this knowledge, individuals can make conscious choices that reduce their environmental impact and contribute to a more sustainable world.

Catalyzing International Cooperation: The global nature of climate systems necessitates international cooperation. Climate data sharing transcends borders, enabling countries to collaborate on addressing shared challenges. Agreements like the Paris Agreement are built upon the foundation of reliable and transparent climate data, fostering a collective commitment to combatting climate change.

In a world where the impacts of climate change ripple across societies and ecosystems, climate data collection, analysis, and interpretation stand as beacons of knowledge and hope. They empower us to comprehend the complexities of our planet's climate, inspire meaningful action, and embark on a journey toward a future where the delicate balance of Earth's systems is preserved for generations to come.

Exploration of different climatic zones, including polar, temperate, tropical, and arid regions

Journey Through Earth's Climatic Zones: From Poles to Deserts

Earth's climatic zones are like the diverse chapters of a captivating story, each with its own unique characters, settings, and climatic experiences. As we traverse the globe, we encounter a symphony of climates, from the frozen expanses of the poles to the sun-drenched landscapes of the tropics, each offering a glimpse into the intricate balance between geography, atmosphere, and life.

Polar Climates: The Frozen Frontiers At the polar extremes lie Earth's polar climates – the Arctic in the north and Antarctica in the south. These frigid realms are defined by icy landscapes, long periods of darkness or daylight, and the remarkable adaptations of species to survive in extreme cold. Polar ice sheets and glaciers shape these regions, playing a crucial role in regulating sea levels. As climate change accelerates, these regions become sentinel zones, offering insights into the impacts of warming on global ecosystems.

Temperate Climates: The Balancing Act Between the poles and the equator, temperate climates flourish. These zones experience four distinct seasons – spring, summer, autumn, and winter – each with its own weather patterns. These regions are characterized by a balance between warm and cold, with moderate temperatures that foster diverse ecosystems. Deciduous forests, grasslands, and agricultural landscapes thrive in temperate climates, where flora and fauna have evolved to adapt to changing seasons.

Tropical Climates: Nature's Abundance Cloaked near the equator, tropical climates bask in warmth and sunlight year-round. The tropics are marked by lush rainforests, vibrant coral reefs, and an abundance of biodiversity. These regions receive consistent solar energy, driving intense convection and rainfall. While tropical paradises captivate with their beauty, they also face challenges like deforestation, habitat loss, and vulnerability to extreme weather events like hurricanes and monsoons.

Arid Climates: Land of Scarcity In arid climates, water becomes a precious resource. Deserts and semi-arid regions dot the landscape, characterized by low rainfall, high temperatures, and vast expanses of sand and rock. These environments are often inhospitable to most life forms, but they are not devoid of biodiversity. Adapted flora and fauna, such as cacti and camels, thrive in the harsh conditions. Arid zones also hold historical significance, as ancient civilizations have flourished along river valleys in these regions.

Exploring the In-Between: Subtropical and Alpine Zones Between the tropics and temperate regions lie subtropical zones, which experience warm temperatures and distinct wet and dry seasons. These regions often house unique ecosystems and are critical for agricultural production. Alpine climates, on the other hand, exist at high altitudes, where temperatures drop as elevation increases. These areas are home to rugged mountains, glacial features, and ecosystems that have adapted to extreme cold.

Harmony and Vulnerability: The Interplay of Life and Climate Earth's climatic zones are not static; they interact, shift, and respond to global climate patterns. The interconnectedness of these zones creates a delicate balance that sustains life on our planet. However, these zones are also vulnerable to the impacts of climate change. As temperatures rise, glaciers melt, sea levels fluctuate, and ecosystems transform, these climatic zones become windows through which we witness the profound consequences of our actions on Earth's delicate balance.

A Global Tale of Adaptation and Change As we venture through polar, temperate, tropical, and arid climates, we witness the tapestry of Earth's diversity and resilience. Each climatic zone tells a story of adaptation, struggle, and beauty, reflecting the intricate connections between geography, atmospheric patterns, and the life that thrives within them. Through exploration, understanding, and stewardship, we can embrace the responsibility to protect and preserve these climatic zones for current and future generations.

Factors influencing variations in temperature, precipitation, and atmospheric conditions

Unraveling the Complex Web: Factors Shaping Temperature, Precipitation, and Atmospheric Conditions

The intricate ballet of temperature, precipitation, and atmospheric conditions that shape our world is guided by a symphony of factors, each playing a distinct role in the dynamic dance of Earth's climate. From the movement of air masses to the geography of the land, these influences interweave to create the diverse climates we experience across the globe.

Solar Influence: The Ultimate Source At the heart of Earth's climate system is the radiant energy of the sun. Solar radiation warms the atmosphere, drives evaporation, and fuels atmospheric circulation. The angle and intensity of sunlight vary with latitude and season, giving rise to temperature gradients that set the stage for climate diversity. As Earth orbits the sun, these variations drive the cycles of day and night, the changing of seasons, and the dynamic interplay of temperature and precipitation.

Geographic Features: The Shape of the Land The configuration of continents, oceans, mountains, and valleys has a profound impact on local climate patterns. Mountains intercept moist air masses, triggering orographic rainfall on windward slopes and creating rain shadows on the leeward side. Large bodies of water, like oceans, moderate temperature extremes, creating maritime climates with relatively stable conditions. The layout of land masses also guides the pathways of prevailing winds, influencing temperature and precipitation patterns.

Atmospheric Circulation: The Global Wind Symphony Earth's rotation and the differential heating of its surface lead to the formation of atmospheric circulation patterns. The Hadley, Ferrel, and Polar cells shape the distribution of temperature and precipitation. Trade winds, westerlies, and polar easterlies drive air masses across the globe, carrying moisture and influencing regional weather patterns. The Coriolis effect, a result of Earth's rotation, adds an element of deflection to these winds, contributing to the complexity of atmospheric dynamics.

Ocean Currents: Temperature Regulators Oceans, with their vast thermal inertia, play a vital role in regulating global climate patterns. Ocean currents distribute heat around the planet, affecting both nearby coastlines and distant regions. Warm ocean currents can raise temperatures along coastal areas, while cold currents can lead to cooler, drier conditions. The El Niño-Southern Oscillation (ENSO) phenomenon, driven by oceanic and atmospheric interactions in the Pacific, has far-reaching effects on global climate variability.

Elevation: The Vertical Dimension Altitude or elevation plays a crucial role in shaping local climate. As air rises in elevation, it expands and cools, leading to temperature decreases with height. This lapse rate can create distinctive temperature zones on mountains, with snow-capped peaks at higher elevations and warmer conditions at lower altitudes. In this way, elevation contributes to the formation of diverse microclimates within a single region.

Human Influence: A Modern Force In recent decades, human activities have emerged as a significant factor influencing climate. The release of greenhouse gases through burning fossil fuels has led to the enhanced greenhouse effect, causing global temperatures to rise. Urbanization and deforestation can create localized heat islands, altering temperature patterns. Pollution and land use changes can also impact precipitation patterns and air quality.

Interconnected Complexity: Navigating a Dynamic System
These factors are not isolated; they are intricately interconnected, creating a web of interactions that give rise to the complexity of Earth's climate. Climate models seek to simulate these intricate relationships, offering insight into potential future scenarios under various conditions. As we continue to explore and understand these influences, we gain a deeper appreciation for the intricate beauty of our planet's climate and the delicate balance that sustains life.

Ecosystems and biodiversity
unique to each climatic zone

Nature's Masterpieces: Ecosystems and Biodiversity Across Climatic Zones

Earth's diverse climatic zones are not only characterized by their unique weather patterns and atmospheric conditions, but also by the remarkable ecosystems and biodiversity that flourish within them. From the frosty tundras of the polar regions to the lush rainforests of the tropics, each climatic zone hosts a tapestry of life uniquely adapted to its surroundings.

Polar Climates: The Resilient Survivors In the extreme cold of polar regions, life has carved out niches of survival. In the Arctic, hardy plants like mosses and lichens cling to rocky landscapes, providing sustenance for grazing animals like reindeer and musk oxen. In Antarctica, where temperatures plunge to record lows, seals, penguins, and seabirds thrive in the surrounding Southern Ocean. These regions, though harsh, support a delicate balance of life that has evolved to withstand the challenges of cold, darkness, and isolation.

Temperate Climates: Seasons of Diversity Temperate climates are characterized by their distinct seasons, creating habitats that are ever-changing. Deciduous forests burst with vibrant colors in autumn and offer shelter to a diverse array of animals, from squirrels to deer. Grasslands, with their open spaces and hardy vegetation, are home to roaming herds of herbivores like bison and antelope. Wetlands, lakes, and rivers teem with aquatic life, while birds of all feathers migrate to and from these regions, adapting to the changing conditions.

Tropical Climates: Nature's Cornucopia In the tropics, a dazzling array of life flourishes year-round. Rainforests, with their dense canopies and rich undergrowth, are biodiversity hotspots, housing countless species of plants, insects, amphibians, and mammals. Coral reefs, submerged ecosystems of unparalleled beauty, support a kaleidoscope of marine life. From the Amazon to the Congo, these regions are a testament to the harmony of life in environments where sunlight, warmth, and moisture are abundant.

Arid Climates: Life's Ingenious Strategies Deserts and arid regions might seem inhospitable, but they are home to species that have evolved remarkable adaptations. Cacti store water within their fleshy stems, and camels can go days without drinking, relying on their ability to conserve water. Desert animals are often nocturnal, seeking refuge from the searing heat of the day. Plant life in these regions often features drought-resistant leaves and stems designed to minimize water loss.

Subtropical and Alpine Zones: Niche Adaptations Subtropical regions, situated between the tropics and temperate zones, host unique ecosystems that adapt to a mix of conditions. Mediterranean climates, for example, are characterized by dry summers and mild, wet winters. Chaparral vegetation thrives in these areas, providing habitat for a variety of wildlife. Alpine zones, found at high altitudes, feature tough, cold-resistant plants and animals that have adapted to survive in the thin air and harsh conditions.

Caring for Our Planet's Ecological Mosaic The intricate dance of ecosystems and biodiversity across climatic zones underscores the interconnectedness of life on Earth. Each species, no matter how small or seemingly insignificant, plays a role in maintaining the delicate balance of these ecosystems. As we navigate the challenges of climate change, human impact, and habitat loss, it becomes ever more crucial to safeguard the diverse life that inhabits these climatic zones. By preserving biodiversity, we ensure the resilience and sustainability of our planet's natural wonders for generations to come.

Atmospheric circulation patterns and their impact on weather and climate

The Winds of Change: Atmospheric Circulation's Influence on Weather and Climate

The rhythmic dance of atmospheric circulation patterns orchestrates the symphony of weather and climate across the globe. From the gentle breezes of trade winds to the dramatic clashes of air masses, these patterns sculpt our world's climate zones, determine precipitation levels, and dictate the ebb and flow of weather phenomena. Understanding their mechanics illuminates the intricate interplay between Earth's atmosphere, oceans, and land.

Hadley Cells: Equatorial Marvels At the equator, the sun's direct rays warm the surface, causing moist air to rise and create a low-pressure zone. This ascent of air generates the Hadley cells, large-scale atmospheric circulation patterns. As the air reaches higher altitudes, it cools, condenses, and forms clouds, leading to abundant rainfall. The trade winds, surface winds blowing from east to west, emerge from this circulation, influencing weather in tropical regions and shaping climates near the equator.

Ferrel Cells: Mid-Latitude Mixers Between the Hadley cells and polar cells lie the Ferrel cells, responsible for mixing air masses and creating the prevailing westerlies. These cells are characterized by the movement of air from high-pressure regions at the surface to low-pressure regions aloft. This causes surface winds to flow from west to east in mid-latitudes, affecting weather patterns in areas such as North America and Europe. The

Ferrel cells play a role in transporting heat and moisture across the globe.

Polar Cells: Cold Poles and Cold Poles At the poles, cold air descends, creating high-pressure zones. The polar cells drive surface winds called polar easterlies, flowing from the poles to the mid-latitudes. These winds interact with the prevailing westerlies and influence the movement of weather systems. The polar cells contribute to the formation of polar climates and have a significant impact on the distribution of ice and snow.

Jet Streams: High-Speed Highways Above the troposphere, jet streams are powerful bands of fast-moving winds. The polar jet stream and the subtropical jet stream have a profound influence on weather and climate. These high-speed highways guide weather systems, steering storms, and influencing the movement of air masses. The jet streams are not only crucial for understanding weather dynamics but also for aviation, as they can accelerate or decelerate flights traveling in their direction.

Impact on Weather and Climate: Atmospheric circulation patterns are the driving forces behind weather and climate phenomena. They play a critical role in:

1. **Temperature Distribution:** Circulation patterns redistribute heat, leading to temperature variations from equator to poles and influencing the development of different climate zones.
2. **Precipitation Patterns:** The rising and sinking of air masses create regions of high and low pressure, which determine areas of intense rainfall and desertification.
3. **Storm Formation:** The interplay of air masses and jet streams is a key factor in the formation and movement of storms, including hurricanes, typhoons, and tornadoes.
4. **Ocean Currents:** Atmospheric circulation interacts with ocean currents, influencing sea surface temperatures

and the distribution of marine life.

5. **Global Climate Patterns:** The circulation patterns are a significant driver of global climate patterns, affecting everything from monsoons to El Niño and La Niña events.

A Delicate Balance: Impact on Our Planet The harmonious dance of atmospheric circulation patterns shapes our world's climate and weather systems. Yet, this balance can be disrupted by external factors, including human-induced climate change. As we continue to study and monitor these patterns, we gain insights into how they may evolve in response to a changing climate, reminding us of the intricate connections that bind Earth's atmospheric dynamics with the future of our planet's weather and climate.

Ocean currents, their role in heat distribution, and their influence on regional climates

Ocean Currents: Nature's Global Heat Distributors

Beneath the surface of Earth's oceans, an intricate network of currents navigates a journey that spans continents and climate zones. These currents are not only conduits of marine life and nutrients but also play a vital role in distributing heat around the planet. Their movement shapes regional climates, influences weather patterns, and contributes to the delicate balance of Earth's climate system.

The Great Conveyor Belt: Thermohaline Circulation At the heart of ocean currents lies the thermohaline circulation, a vast global system driven by temperature and salinity differences. Cold, dense water sinks in polar regions, while warmer, less dense water rises in the tropics. This movement creates a conveyor belt-like circulation that extends from the deep ocean to the surface and back again, transporting heat and regulating the distribution of warmth across the globe.

Equatorial Currents: Tropical Heat Redistribution Equatorial currents, such as the warm North Equatorial Current and South Equatorial Current, play a critical role in redistributing tropical heat. These currents flow along the equator, carrying warm water from the tropics toward the poles. As warm water moves poleward, it releases heat into the atmosphere, influencing regional climate patterns and contributing to the formation of weather systems.

Gulf Stream and Kuroshio Current: Warmth Across Latitudes
The Gulf Stream in the Atlantic Ocean and the Kuroshio Current in the Pacific Ocean are prominent examples of warm ocean currents that flow from low to high latitudes. These currents carry warm waters from tropical regions to higher latitudes, significantly impacting the climate of adjacent land areas. The Gulf Stream, for instance, helps moderate temperatures along the eastern coast of North America and northwestern Europe, leading to milder winters and cooler summers.

Cold Ocean Currents: Cooling Effects Cold ocean currents, like the California Current along the west coast of North America and the Canary Current along the northwest coast of Africa, have a cooling influence on adjacent regions. These currents originate from polar or cold waters and move toward lower latitudes. As they flow along the coastline, they can bring cooler air and lower sea surface temperatures, impacting local climates and marine ecosystems.

El Niño and La Niña: Disrupting the Balance Ocean currents play a central role in the El Niño-Southern Oscillation (ENSO), a climate phenomenon that disrupts normal weather patterns in the tropical Pacific Ocean. During El Niño events, warm waters from the western Pacific flow eastward, altering atmospheric conditions and influencing global climate patterns. Conversely, La Niña events involve cooler waters in the central and eastern Pacific, triggering different climate impacts.

Regional Climates and Weather Patterns: The influence of ocean currents on regional climates and weather patterns is profound:

1. **Coastal Climates:** Ocean currents moderate coastal climates, leading to milder temperatures and influencing precipitation patterns. Coastal areas often experience more stable temperatures due to the thermal inertia of ocean waters.
2. **Rainfall Distribution:** Warm ocean currents can enhance evaporation, leading to higher moisture levels

in the atmosphere and potentially contributing to increased rainfall in coastal and nearby regions.

3. **Hurricane Formation:** Warm ocean waters provide the energy necessary for hurricane formation and intensification. The Gulf Stream, for example, can fuel hurricanes as they travel across the Atlantic Ocean.

4. **Marine Life:** Ocean currents transport nutrients and influence marine ecosystems, impacting the distribution of species and the productivity of coastal and open ocean areas.

A Global Symphony: Connecting Climates Ocean currents form an intricate symphony that connects climates, continents, and ecosystems. Their ability to distribute heat across the planet helps shape regional climate patterns, influence weather events, and sustain marine life. As we continue to study and understand the intricate dynamics of ocean currents, we gain a deeper appreciation for their role in the harmonious interplay of Earth's climate system.

Interactions between land, sea, and atmosphere that shape climate characteristics

The Dance of Elements: Interactions Shaping Climate Characteristics

The interconnectedness of land, sea, and atmosphere orchestrates a dynamic symphony of interactions that give rise to the diverse climate characteristics observed across the globe. From the temperature of the air to the moisture in the soil, these elements collaborate to shape the intricate patterns of weather and climate that define our planet.

Land-Water Contrast: Thermal Dynamics The contrasting properties of land and water create a fundamental influence on climate characteristics. Land heats up and cools down more quickly than water. As a result, coastal areas experience milder and more stable temperatures compared to inland regions. During the day, land heats up more rapidly, leading to the development of local breezes as air moves from the cooler sea to the warmer land. At night, the opposite occurs, with land cooling more rapidly and sea breezes moving inland.

Oceanic Influence: Maritime Moderation The oceans play a significant role in moderating temperatures of nearby land areas. Coastal regions experience a more stable climate due to the thermal inertia of water. Ocean currents, influenced by atmospheric circulation patterns, transport heat across vast distances, influencing regional climates. For instance, the warm Gulf Stream has a profound impact on the climate of eastern North America and northwestern Europe, keeping winters milder and summers cooler.

Latitudinal Variations: Solar Energy Distribution The Earth's spherical shape leads to variations in solar energy received across different latitudes. Near the equator, the sun's direct rays create warmer temperatures, while higher latitudes receive less direct sunlight and experience cooler conditions. This latitudinal variation is a fundamental driver of climate zones, from the tropical heat of the equator to the cold of the polar regions.

Topography and Elevation: Vertical Climate Zones The elevation of landforms creates distinct vertical climate zones. As air rises in elevation, it cools and condenses, leading to the formation of clouds and precipitation. Mountain ranges can act as barriers, causing moist air to rise and release moisture on the windward side (orographic rainfall), resulting in lush vegetation. On the leeward side, rain shadows form, creating drier conditions.

Atmospheric Circulation: Wind and Weather The movement of air masses driven by atmospheric circulation patterns is a central element in shaping climate characteristics. Trade winds, westerlies, and polar easterlies transport air masses and moisture, influencing weather patterns and precipitation. The Hadley, Ferrel, and Polar cells create a framework for understanding how air circulates around the globe and how this movement shapes regional climates.

Interaction with Vegetation: Land Cover Influence Vegetation, from lush forests to arid deserts, plays a role in shaping local climates. Forests release moisture through a process called transpiration, which can lead to increased humidity and cloud formation. In contrast, arid regions with sparse vegetation experience less transpiration, resulting in drier conditions. Changes in land cover due to deforestation or urbanization can alter local climate patterns.

Climate Feedback Loops: Complex Interactions These interactions are not isolated but are part of complex feedback loops. For example, as temperatures rise due to climate change, melting ice and warmer oceans can impact atmospheric circulation patterns, influencing weather systems and climate zones. These interactions highlight the importance of understanding the delicate balance between land, sea, and atmosphere.

A Holistic Portrait: Earth's Climate Mosaic The interactions between land, sea, and atmosphere create a diverse and intricate mosaic of climate characteristics. From the majestic peaks of mountains to the tranquil embrace of oceans, each element contributes its unique voice to the composition of Earth's climate. As we explore the tapestry of interactions, we gain a deeper appreciation for the delicate equilibrium that sustains our planet's climate and the interconnectedness that shapes its complexity.

Uncovering historical climate data sources, such as ice cores, tree rings, and sediment layers

Chronicles of the Past: Unveiling Historical Climate Data

To understand the ever-changing climate of our planet, scientists turn to a treasure trove of historical climate data sources that span millennia. These records, hidden within the layers of ice, the growth rings of trees, and the sediments of ancient lakes and oceans, provide a window into the Earth's climatic history. By deciphering these natural archives, researchers gain insights into past climate variations, helping to unravel the mysteries of our planet's climatic past.

Ice Cores: Time Capsules of Frozen History Ice cores extracted from polar ice sheets and glaciers offer a glimpse into the climate's deep past. Each layer of ice represents a snapshot of atmospheric conditions at the time it formed. By analyzing the air trapped in bubbles within the ice, scientists can reconstruct past atmospheric composition, including greenhouse gas concentrations. Isotopic ratios in the ice reveal temperature variations, with thicker layers indicating warmer periods and vice versa. Ice cores from Antarctica, Greenland, and other regions provide a continuous record stretching back hundreds of thousands of years.

Tree Rings: Nature's Annals of Growth Trees bear a record of the climate in the patterns of their growth rings. Each year, a new ring forms, reflecting the interplay of temperature, precipitation, and other environmental factors. Wide rings indicate favorable conditions for growth, while narrow rings indicate stress. By cross-dating tree ring patterns from different trees, researchers create master chronologies that extend back centuries, offering insights into regional and global climate variability.

Sediment Layers: Preserving Clues in the Earth's Archives Sediments accumulated in lakes, oceans, and riverbeds hold a wealth of information about past climate conditions. Microscopic fossils, pollen grains, and geochemical markers found in sediment layers can provide clues about temperature, precipitation, and sea level changes. By drilling cores from the depths of these sediment deposits, scientists can reconstruct the environmental history of a particular region, uncovering periods of drought, flooding, and other climatic events.

Cave Speleothems: Nature's Timekeepers Underground Stalagmites and stalactites in caves, known as speleothems, grow layer by layer as mineral-rich water drips and evaporates. These formations capture climate information in the chemical composition of their layers. Oxygen isotope ratios in speleothems reflect past temperature and rainfall patterns. By studying these formations, researchers can reconstruct paleoclimate records that date back thousands of years.

Historical Records: Insights from Human Narratives In addition to natural archives, historical records created by human societies offer glimpses into past climate variations. Historical documents, diaries, ship logs, and records of harvests provide indirect clues about climate conditions and their impact on societies. These records can help bridge the gap between natural archives and human experiences, providing a more comprehensive understanding of past climate variability.

Decoding the Past: Insights for the Future These historical climate data sources collectively serve as time capsules that unlock the Earth's climatic history. By piecing together information from ice cores, tree rings, sediment layers, and more, scientists create a rich tapestry of past climate variability.

This knowledge not only deepens our understanding of natural climate cycles but also provides essential context for evaluating the unprecedented changes occurring in the modern era. As we look to the future, these historical records remind us of the Earth's resilience, while also urging us to consider the implications of our actions on the climate of tomorrow.

Using proxies to reconstruct past climates and understand long-term climate trends

Proxies: The Silent Witnesses of Past Climates

In the intricate puzzle of Earth's climatic history, proxies are the invaluable clues that provide insights into the tapestry of past climates. These indirect sources of information, found in natural archives such as ice, sediment, trees, and more, allow scientists to reconstruct the conditions of ancient environments. By deciphering these proxies, researchers gain a deeper understanding of long-term climate trends, offering a glimpse into the Earth's climatic past.

Ice Cores: Frozen Time Capsules Ice cores, taken from polar ice sheets and glaciers, encapsulate a wealth of information about past climates. Within the layers of ice, trapped air bubbles hold a record of past atmospheric composition. Isotopic ratios of oxygen and hydrogen in the ice reveal temperature variations over time. As researchers drill deeper into the ice, they delve farther into history, piecing together a narrative of ancient climates and the ebb and flow of ice ages.

Tree Rings: Nature's Growth Rings of Time Trees silently chronicle the passing years through their growth rings. These rings not only reflect the changing seasons but also hold information about temperature, precipitation, and environmental stress. Wide rings suggest favorable conditions, while narrow rings indicate challenging times. By analyzing tree ring patterns across regions, scientists construct a mosaic of past climate variability, offering insights into droughts, heatwaves, and other phenomena.

Sediment Layers: Earth's Sedimentary Archives Sediments in lakes, oceans, and riverbeds serve as archives of past climates. Within these layers lie tiny fossils, pollen grains, and mineralogical clues. The abundance of specific organisms can hint at temperature and moisture conditions. Changes in sediment composition, such as the presence of glacial debris, provide a chronicle of ice ages. As sediment cores are meticulously analyzed, a vivid picture of Earth's climatic history emerges.

Cave Deposits: Hidden Treasures Underground Caves hold secrets in their formations known as speleothems, such as stalagmites and stalactites. These mineral deposits accumulate over time as water drips and evaporates. The isotopic composition of oxygen in these formations reveals information about temperature and rainfall patterns. As scientists study speleothems, they journey into the depths of Earth's climatic past, uncovering the delicate balance of nature.

Proxy Networks: A Symphony of Clues What makes proxies truly powerful is their ability to form a network. By combining multiple proxies, researchers can paint a comprehensive picture of past climates. For example, the correlation between ice core data, tree ring records, and sediment layers offers a multidimensional perspective on ancient environments. These proxy networks reveal patterns, anomalies, and trends that help scientists piece together the puzzle of Earth's long-term climate dynamics.

Gaining Insight for Our Future The study of proxies is not merely a journey into the past—it holds lessons for our future. By understanding the Earth's natural variability and the factors that have shaped its climates over millions of years, scientists can better interpret the unprecedented changes occurring in the modern era. Proxies remind us that climate is a complex interplay of various factors, urging us to approach climate science with humility and a commitment to understanding and stewarding our planet's delicate balance.

Insights into natural climate variability and patterns over centuries and millennia

Time's Echoes: Insights into Centuries and Millennia of Climate Variability

The rhythmic heartbeat of Earth's climate spans centuries and millennia, weaving a complex tapestry of natural variability. Through the lenses of historical records, proxy data, and scientific inquiry, we catch glimpses of the planet's climatic ebb and flow. These insights not only illuminate the past but also provide essential context for comprehending the present and contemplating the future of our planet's climate.

Proxy Perspectives: Unveiling Ancient Secrets Proxies—hidden within ice, trees, sediments, and caves—serve as portals to bygone eras. Analyzing their patterns and records unveils a mosaic of natural climate variability. Ice cores whisper tales of glacial advances and retreats, tree rings recount periods of warmth and frost, sediment layers narrate stories of flooding and drought, and cave formations echo the eons in their isotopic signatures. These proxies, when interwoven, reveal cycles of climate oscillations, from short-term fluctuations to long-term trends.

Milankovitch Cycles: Earth's Orbital Waltz One of nature's choreographed dances is the Milankovitch cycles—variations in Earth's orbit, axial tilt, and precession. Over millennia, these cycles influence the amount and distribution of solar energy reaching the planet's surface. They play a role in shaping ice ages and interglacial periods. The cyclical nature of these changes explains much of the natural climate variability observed over tens of thousands of years.

Medieval Warm Period and Little Ice Age: Glimpses into Anomalies The Medieval Warm Period (roughly 950-1250 CE) and the subsequent Little Ice Age (roughly 1350-1850 CE) punctuate history with departures from the norm. The Medieval Warm Period brought relatively mild temperatures, while the Little Ice Age saw colder conditions. These periods, revealed through proxy records and historical accounts, provide insight into the natural oscillations of climate and the effects of factors like solar variability and volcanic activity.

Pacific Decadal Oscillation and Atlantic Multidecadal Oscillation: Deciphering Complex Oscillations Patterns of natural climate variability also emerge on shorter timescales. Oscillations like the Pacific Decadal Oscillation (PDO) and the Atlantic Multidecadal Oscillation (AMO) bring about shifts in sea surface temperatures and atmospheric circulation. These oscillations, lasting several decades, influence regional climates, including the frequency of droughts, floods, and storm activity.

Long View on Natural Climate Change: Lessons for Today Studying natural climate variability over centuries and millennia offers crucial lessons for understanding our current climate reality. It underscores the inherent fluctuations that exist in Earth's climate system. These fluctuations, driven by factors such as solar variability, volcanic activity, and natural internal processes, remind us that change is an inherent part of Earth's climate story.

Context for Anthropogenic Change: A Nuanced Perspective As we navigate the complexities of anthropogenic climate change, insights into natural climate variability provide a backdrop against which to assess the scale and pace of current changes. They offer a nuanced perspective, highlighting the need to distinguish between natural variations and human-induced changes. This distinction guides our efforts to mitigate and adapt to the rapid changes occurring in the modern era.

Preserving the Rhythm: Navigating the Future The study of climate variability across centuries and millennia beckons us to preserve the rhythm of our planet's climate system. As we ponder the echoes of the past, we are called to take responsible action to ensure a balanced and resilient future. By harmonizing our actions with the natural pulse of Earth's climate, we contribute to a symphony that transcends generations and safeguards the delicate balance of our world.

The concept of climate change and its driving factors, including greenhouse gas emissions

Climate Change Unveiled: Understanding its Drivers and Dynamics

Climate change is a profound transformation that has gripped our planet, reshaping the very fabric of its ecosystems and weather patterns. This transformation, driven by a complex interplay of natural and human factors, stands as one of the most pressing challenges of our time. To comprehend this phenomenon, we must peel back its layers and explore the driving forces that propel it, chief among them being the emissions of greenhouse gases.

Greenhouse Effect: Nature's Blanket At the heart of the matter lies the greenhouse effect, a natural phenomenon crucial for maintaining Earth's habitable conditions. Certain gases in the atmosphere, such as carbon dioxide (CO_2), methane (CH_4), and water vapor, act like a thermal blanket, allowing sunlight to enter but trapping heat that would otherwise escape into space. This natural process keeps our planet's temperature within a livable range.

Anthropogenic Influence: The Human Fingerprint However, the intensification of the greenhouse effect in recent decades is largely attributed to human activities. The burning of fossil fuels —coal, oil, and natural gas—releases vast amounts of CO_2 into the atmosphere. Deforestation and land use changes disrupt the planet's natural carbon balance. Industrial processes and agriculture emit methane and other potent greenhouse gases. The result is an accumulation of these gases, thickening the thermal blanket and leading to a rise in global temperatures—a phenomenon commonly referred to as global warming.

Carbon Footprints: Tracing Human Impact Measuring the human impact on the climate is done through carbon footprints— the total amount of greenhouse gas emissions, expressed in units of CO2-equivalent, associated with an individual, organization, or activity. From transportation to energy consumption, each aspect of modern life contributes to these footprints, leaving behind a trail of carbon emissions that drive climate change.

Positive Feedback Loops: Amplifying Effects Climate change isn't confined to isolated temperature increases. It sets in motion a cascade of effects known as positive feedback loops. As ice melts due to warming, the Earth's albedo (reflectivity) decreases, absorbing more heat and accelerating ice melt further. Thawing permafrost releases methane, a potent greenhouse gas, intensifying the warming effect. These amplifying feedbacks create a chain reaction that magnifies the impacts of climate change.

Impacts and Consequences: A Global Story The consequences of climate change are far-reaching and diverse. More frequent and intense heatwaves, altered precipitation patterns, sea level rise, and shifting ecosystems are just a few threads in this intricate tapestry. Communities, particularly vulnerable populations, face challenges related to water scarcity, food security, and displacement due to sea level rise. Biodiversity loss and disruptions to agriculture, infrastructure, and economies are woven into this narrative.

Mitigation and Adaptation: Navigating the Future Addressing climate change requires a multi-pronged approach. Mitigation efforts aim to reduce greenhouse gas emissions through transitioning to renewable energy sources, increasing energy efficiency, and reforestation, among other measures. Adaptation strategies focus on preparing communities for the changes already set in motion, building resilience to withstand the impacts of a changing climate.

A Call to Action: Shaping a Sustainable Tomorrow Understanding the concept of climate change is not a mere intellectual exercise—it's a call to action. It's a call to embrace clean energy, promote sustainable practices, and make choices that mitigate our impact on the planet. As we unravel the driving factors behind climate change, we also unveil our responsibility to shape a future that preserves the delicate balance of Earth's ecosystems and secures the well-being of present and future generations.

The role of human activities in accelerating global warming and altering the climate

Human Hands, Planetary Consequences: Accelerating Global Warming

The pages of Earth's climatic history are now being written with a distinct ink—a potent blend of human activities that has accelerated global warming and set in motion a cascade of climate alterations. From the burning of fossil fuels to deforestation, our actions have cast a profound influence on the delicate balance of the planet's climate system. Unveiling the role of human activities in this intricate tale is essential to understanding the urgency of addressing this global challenge.

Fossil Fuel Emissions: A Heat-Generating Symphony The burning of fossil fuels—coal, oil, and natural gas—releases a symphony of greenhouse gases into the atmosphere. Carbon dioxide (CO_2) takes center stage, acting as the conductor of global warming. As fossil fuels power our industries, vehicles, and homes, CO_2 concentrations rise, thickening the atmospheric blanket that traps heat. This human-made increase in CO_2 levels is directly linked to the significant rise in global temperatures observed over recent decades.

Deforestation and Land Use Changes: A Disrupted Carbon Balance Forests, nature's carbon reservoirs, play a pivotal role in the planet's carbon cycle. However, deforestation and land use changes have disrupted this delicate balance. Trees that once absorbed CO_2 are felled, releasing stored carbon back into the atmosphere. The loss of these carbon sinks intensifies the greenhouse effect and amplifies global warming.

Agriculture and Livestock: Methane and Nitrous Oxide Duets Human activities in agriculture contribute their own notes to the composition of climate change. Livestock, particularly cattle, produce methane—a potent greenhouse gas—during digestion. Additionally, agricultural practices that involve fertilizers release nitrous oxide, another potent contributor to global warming. These emissions from agriculture create harmonies of warming effects that reverberate across the atmosphere.

Industrial Processes: Singing a Chorus of Emissions Industrial processes, essential to modern life, also contribute to the choir of emissions. Manufacturing, cement production, and chemical industries release greenhouse gases such as methane, nitrous oxide, and fluorinated gases into the air. These emissions further amplify the greenhouse effect, nudging global temperatures upward.

Urbanization and Transportation: Cities in the Climate Spotlight As populations concentrate in urban areas, the demand for energy and resources skyrockets. Urbanization leads to increased energy consumption, emissions from vehicles, and the heat island effect, where urban areas are hotter than surrounding rural areas. The combined impact of urbanization and transportation adds to the crescendo of warming influences.

Feedback Loops: Intensifying the Melody Human-induced warming also triggers feedback loops that amplify its effects. As ice and snow melt due to higher temperatures, the Earth's albedo decreases, causing more sunlight to be absorbed and accelerating warming. Thawing permafrost releases methane, a greenhouse gas, into the atmosphere, contributing to further warming. These feedback loops create a cycle of amplification that heightens the impact of human activities on the climate.

A Concerted Response: Composing a Sustainable Future Recognizing the role of human activities in accelerating global warming is a call to action—a call to harmonize our efforts for a sustainable future. Transitioning to clean and renewable energy sources, adopting sustainable agricultural practices, reforestation, and promoting energy efficiency are key notes in this symphony of solutions. The role we play in altering the climate underscores our shared responsibility to shape a world where the melody of human progress resonates in harmony with the rhythms of nature.

Impacts of climate change on ecosystems, sea levels, and extreme weather events

Unraveling the Tapestry: Climate Change's Impact on Earth's Fabric

As the threads of our planet's climate weave a new narrative, the story unfolds with dramatic shifts that reverberate across ecosystems, shorelines, and skies. Climate change, fueled by human activities, has unleashed a series of interconnected impacts, reshaping the very essence of Earth's landscapes. From altered ecosystems to rising sea levels and intensified extreme weather events, the effects of this global transformation are written into the fabric of our world.

Ecosystems in Flux: A Symphony of Change Ecosystems, finely tuned to specific climate conditions, are undergoing a symphony of alterations. Rising temperatures shift the ranges of plants and animals, pushing species toward cooler habitats or even leading to extinction. Migratory patterns are disrupted, affecting pollination and predator-prey relationships. Coral reefs, fragile marine ecosystems, suffer from bleaching events due to warmer ocean temperatures. The delicate balance of ecosystems is disrupted, threatening biodiversity and the services they provide.

Rising Tides, Shifting Landscapes: Sea Level Rise Unveiled One of the most tangible impacts of climate change is the rise in sea levels. Melting glaciers and ice sheets, coupled with the thermal expansion of seawater, contribute to this phenomenon. Rising seas encroach upon coastlines, eroding shorelines, submerging low-lying areas, and threatening coastal communities. Saltwater intrusion affects freshwater sources and disrupts ecosystems in coastal zones. The implications for human societies, economies, and the delicate balance of coastal ecosystems are profound.

Fury Amplified: Intensified Extreme Weather Events Extreme weather events—heatwaves, hurricanes, droughts, and floods—are not new, but climate change has amplified their intensity and frequency. Warmer temperatures provide fuel for more intense hurricanes and typhoons, leading to devastating impacts on communities. Heatwaves become more frequent, posing health risks and straining energy resources. Rainfall patterns shift, causing floods in some regions and droughts in others. These events test the resilience of infrastructure, economies, and ecosystems.

Arctic Amplification: Melting Ice and Altered Dynamics The Arctic, a region sensitive to climate change, experiences amplified warming known as Arctic amplification. Melting ice contributes to rising sea levels while altering ocean currents and atmospheric circulation patterns. The loss of sea ice affects the habitats of polar species, from polar bears to seals and whales. The disappearing ice cover also opens up new shipping routes, triggering geopolitical and environmental concerns.

Cascading Effects: Interconnected Impacts What sets climate change apart is its ability to trigger cascading effects. As ecosystems shift, it affects not only species directly but also the human communities that rely on them for food, livelihoods, and cultural practices. Rising sea levels influence coastal economies, infrastructure, and migration patterns. Extreme weather events lead to damage and displacement, straining resources and challenging disaster response capabilities.

Adapting to Change: Navigating the Future Addressing the impacts of climate change requires a multi-faceted approach. Mitigation efforts aim to curb greenhouse gas emissions, slowing the pace of change. Adaptation strategies focus on building resilience, whether through building seawalls, implementing water management techniques, or conserving critical ecosystems. Collaborative international efforts and policy initiatives are essential to address the global nature of climate change.

A Call to Action: Protecting Earth's Legacy The impacts of climate change are not distant scenarios—they are unfolding before our eyes. Our shared responsibility is to embrace the call to protect and restore the balance of our planet's ecosystems, safeguard coastal communities, and fortify societies against the fury of extreme weather events. By understanding the intricacies of these impacts, we empower ourselves to make informed decisions, raise awareness, and work collectively toward a future where the tapestry of Earth's landscapes remains resilient, vibrant, and harmonious.

Exploring paleoclimatology as a window into Earth's climate history

Paleoclimatology: Peering through Earth's Time Portal

The Earth's history, hidden within its geological layers, whispers tales of ancient climates through the science of paleoclimatology. This remarkable discipline allows us to unlock the secrets of bygone eras, providing a panoramic view of how our planet's climate has ebbed and flowed over millions of years. With proxies as its guide, paleoclimatology opens a window into Earth's climate history, offering valuable insights into the past and illuminating the trajectory of our future.

Proxy Portraits: Voices from the Past Proxies—nature's time capsules—are the instruments of paleoclimatology. These indirect records, found in ice cores, sediment layers, tree rings, and other natural archives, preserve traces of past climates. By deciphering the information encoded within these proxies, scientists reconstruct temperature patterns, precipitation variations, and atmospheric conditions that span epochs.

Ice Cores: Frozen Chronicles The ice sheets and glaciers that span the polar regions harbor a treasure trove of climate information. Ice cores drilled from these icy landscapes harbor layers of trapped air bubbles, each one a snapshot of the atmosphere's composition at a specific point in time. Isotopic ratios within the ice reveal temperature changes, while impurities offer insights into past volcanic eruptions and air quality.

Sediment Layers: Earth's Memory Bank Sediments that

accumulate in lakes, oceans, and riverbeds serve as archives of climatic change. Within these layers lie the fossilized remains of ancient organisms, pollen grains, and mineral deposits. The composition and arrangement of these sediments unveil shifts in temperature, precipitation, and sea level over time.

Tree Rings: Nature's Yearbook Trees, with their annual growth rings, chronicle the passage of time. Each ring reflects the conditions of a specific year, including temperature, moisture, and sunlight availability. By analyzing the patterns in tree rings, scientists construct a timeline of climate variability that stretches back hundreds or even thousands of years.

Cave Formations: Geological Diaries Caves, adorned with stalagmites and stalactites, harbor mineral formations known as speleothems. These formations grow layer by layer, much like the rings of a tree. The isotopic composition of these layers provides clues about temperature and precipitation variations in the past, offering a glimpse into Earth's climatic history.

Unraveling Earth's Story: Insights and Implications The stories told by paleoclimatology carry profound implications for our understanding of Earth's climate system. By charting the rise and fall of temperatures, the waxing and waning of ice ages, and the rhythms of natural climate variability, we gain a broader context for the changes occurring today. These insights underscore the significance of anthropogenic climate change and offer valuable data for climate models that project future scenarios.

Guiding Our Path: Learning from the Past As we face the challenges of a rapidly changing climate, paleoclimatology serves as a compass, guiding us toward informed decisions and adaptive strategies. By learning from Earth's past, we can better anticipate future changes, plan for resilience, and mitigate the impacts of climate change. The study of paleoclimatology transcends time, providing us with a powerful tool to navigate the complex interplay between human activities and the delicate balance of our planet's climate.

The study of past climate conditions through geological records and proxy data

Journeying Through Time: Decoding Earth's Climate Chronicles

The Earth is a historian, its geological records serving as ancient manuscripts that recount the climatic stories of ages past. Through the diligent work of scientists, the discipline of studying past climate conditions—often referred to as paleoclimatology—unlocks the mysteries of bygone eras. By examining geological records and proxy data, this field weaves together a tapestry of Earth's climatic history, offering invaluable insights into the forces that have shaped our world.

Geological Archives: Layers of Earth's Memory The planet's geological layers, meticulously crafted over millions of years, bear witness to the ebb and flow of climate patterns. Sediments in lakes, oceans, and riverbeds hold a trove of information. With each layer deposited, a new chapter of Earth's climate narrative is inscribed. By analyzing the composition, texture, and arrangement of these sediments, scientists uncover clues about past temperatures, precipitation levels, and even ancient landscapes.

Proxy Data: Echoes of the Past While Earth's geological archives provide the canvas, proxy data are the brushstrokes that paint the picture of past climates. Proxies—indirect indicators of climate conditions—exist in various forms across the natural world. Ice cores, for example, are like frozen time capsules, preserving ancient air bubbles that reveal atmospheric composition and temperatures. Tree rings offer annual snapshots of conditions in their growth patterns. Coral reefs, with their intricate structures, record sea surface temperatures and oceanic changes.

The Ice Core Chronicles: A Glimpse of Ancient Atmospheres In polar ice sheets and glaciers, layers of ice act as the pages of an ancient diary. Each layer, formed by accumulated snow over millennia, captures atmospheric composition and temperature at the time it was deposited. Air bubbles trapped within the ice contain samples of ancient atmospheres, allowing scientists to peer back in time. As ice cores are extracted and analyzed, they unveil the chemical makeup of the past, revealing a history of greenhouse gas levels and climatic conditions.

Tree Rings: Time Travelers in the Forest The growth rings of trees are like time travelers, preserving a record of past growing seasons. By studying the thickness of these rings, scientists can infer climate conditions during each year of a tree's life. Narrow rings might indicate drought years, while wider rings suggest favorable conditions. By examining tree ring patterns across many trees and regions, researchers construct a timeline of past climate variability.

Cave Deposits: Stalagmites and Stalactites Speak Caves are Earth's underground libraries, preserving climatic tales within their mineral formations. Stalagmites and stalactites, formed drip by drip, capture the isotopic composition of the water that created them. By analyzing these formations, scientists can glean information about past temperatures, rainfall patterns, and even ancient vegetation cover.

Unlocking Earth's Secrets: Implications for Today and Tomorrow The study of past climate conditions isn't just about uncovering the past; it's about understanding our present and shaping our future. By reconstructing past climate variability, scientists gain insights into the natural rhythms and variations of the climate system. This knowledge helps contextualize the changes occurring in today's world, including the influence of human activities. It also informs climate models, allowing us to make more accurate predictions about future climate scenarios.

A Roadmap for Resilience: Navigating Forward The study of past climate conditions isn't solely an academic pursuit—it's a roadmap for resilience. By learning from Earth's history, we gain a deeper understanding of the potential impacts of a changing climate. This understanding equips us to adapt, innovate, and develop strategies that can mitigate the challenges posed by ongoing climate change. In the annals of Earth's geological records and proxy data, we find not just stories of the past, but lessons to guide us on a more sustainable and informed journey into the future.

Insights into ancient climates, ice ages, and shifts in atmospheric composition

Echoes of Eons: Unveiling Ancient Climates and Shifting Skies

The Earth's history is a symphony of climatic shifts, orchestrated by the interplay of geological forces and atmospheric dynamics. Through the lens of paleoclimatology, we journey back in time to explore the rich tapestry of ancient climates, witness the rise and retreat of ice ages, and decipher the evolving composition of Earth's atmosphere. These insights not only illuminate the past but offer profound implications for our understanding of the present and the trajectory of our future.

Time Capsules in Ice: Reconstructing Ancient Atmospheres The glaciers and ice sheets that cloak the polar regions hold the secrets of past climates within their frozen cores. As scientists drill into these icy archives, they extract cylindrical records that span hundreds of thousands of years. The air bubbles trapped within these ice cores provide a glimpse into ancient atmospheres. By analyzing the ratios of isotopes and trace gases, researchers deduce temperatures, atmospheric carbon dioxide levels, and even the composition of ancient air.

Ice Ages and Interglacials: Earth's Climate Oscillations The Earth's climate history is punctuated by the rhythmic dance of ice ages and interglacial periods. These cycles, driven by changes in Earth's orbit and axial tilt, have sculpted the planet's landscape over millions of years. During ice ages, vast ice sheets advanced across continents, locking up water and lowering sea levels. Interglacial periods brought warmer temperatures, ice retreat, and the reshaping of ecosystems. Studying these cycles provides insights into the forces that shape Earth's climate over geological timescales.

The Breath of the Past: Shifting Atmospheric Composition Earth's atmosphere is a dynamic canvas, painted with gases that have evolved over eons. Through proxies like ice cores and sediment layers, scientists piece together the puzzle of atmospheric composition. The transition from higher to lower levels of carbon dioxide during ice ages reflects the exchange of carbon between the atmosphere, oceans, and biosphere. Changes in methane levels reveal the influence of wetlands, agriculture, and human activities on the atmosphere's chemical makeup.

A Portrait of Earth's Past: Lessons for Today and Tomorrow The study of ancient climates and shifting atmospheric compositions yields more than just scientific curiosity—it carries essential lessons for the modern world. By understanding the natural variability of Earth's climate, we gain a broader perspective on the changes occurring today. This understanding highlights the role of human activities in accelerating climate change and underscores the need for informed decision-making to mitigate its impacts.

Navigating the Anthropocene: Wisdom from the Ages As we navigate the challenges of the Anthropocene—the epoch defined by human influence on the planet—we draw on the lessons of Earth's ancient climates. The ebb and flow of ice ages, the fluctuations in greenhouse gas concentrations, and the resilience of ecosystems in the face of change all offer insights into our ability to adapt and innovate. By heeding the wisdom of ancient climates, we shape a future that respects the delicate balance of Earth's systems, safeguards biodiversity, and ensures the well-being of generations yet to come.

Strategies for adapting to changing climatic conditions and mitigating risks

Navigating Uncertain Waters: Strategies for Climate Resilience and Mitigation

As the climate landscape continues to evolve, humanity finds itself at a crucial crossroads. The impacts of climate change, driven by both natural forces and human activities, necessitate proactive strategies that foster resilience and mitigate risks. Adapting to changing climatic conditions and curbing the pace of global warming require innovative and coordinated efforts that span governments, industries, communities, and individuals. These strategies are not only a response to a warming world but a blueprint for safeguarding our planet's future.

Climate Resilience: Fortifying Against the Storm Building climate resilience is a cornerstone of adaptation strategies. This involves anticipating and responding to the challenges posed by a changing climate, from more intense heatwaves to rising sea levels. Designing infrastructure that can withstand extreme weather events, implementing flood management systems, and relocating vulnerable communities away from high-risk areas are all part of the resilience equation. By preparing for the inevitable impacts of a warming world, we create a foundation for enduring change.

Ecosystem-based Adaptation: Nature as a Partner Nature has a remarkable ability to adapt to changing conditions, and harnessing its power can be a key component of climate adaptation. Ecosystem-based adaptation involves restoring and conserving natural systems like wetlands, forests, and mangroves to provide natural buffers against climate impacts. These ecosystems regulate water flow, absorb carbon dioxide, and protect against storm surges. By working with nature rather than against it, we enhance our ability to withstand climatic challenges.

Climate-Resilient Agriculture: Sowing Seeds of Change Agriculture, vital for global food security, faces increasing challenges from climate change. Shifting weather patterns, droughts, and changing pest dynamics threaten crop yields and livelihoods. Climate-resilient agricultural practices involve diversifying crops, improving water management, and adopting sustainable farming techniques. Investing in resilient agriculture not only ensures food security but also contributes to carbon sequestration and ecosystem health.

Renewable Energy Revolution: Curbing Carbon Emissions Mitigating climate change hinges on transitioning from fossil fuels to renewable energy sources. Solar, wind, hydroelectric, and geothermal energy offer cleaner alternatives that reduce greenhouse gas emissions. Investing in renewable energy infrastructure, incentivizing energy efficiency, and phasing out coal and oil power plants are crucial steps toward a low-carbon future. This not only curbs global warming but also fosters economic growth and energy independence.

Green Urban Planning: Building Sustainable Cities Cities, home to a majority of the world's population, are hubs of innovation and consumption. Green urban planning integrates sustainable transportation, energy-efficient buildings, and green spaces to create livable and climate-resilient cities. Reducing urban heat islands, promoting public transportation, and adopting energy-saving technologies contribute to both quality of life and carbon reduction.

International Cooperation: Bridging Boundaries for a Global

Solution Climate change transcends borders, demanding international collaboration to tackle its challenges. Global agreements like the Paris Agreement aim to unite nations in their efforts to limit global warming. By committing to emission reduction targets and climate adaptation plans, countries send a powerful message that collective action is essential for safeguarding the planet.

Individual Action: Every Step Matters From policy makers to individuals, every actor plays a role in climate adaptation and mitigation. Individuals can make a difference by adopting sustainable lifestyles, reducing energy consumption, minimizing waste, and advocating for climate-conscious policies. Small choices add up to significant impacts when scaled across communities and nations.

A Future of Promise: Walking the Path Together Adapting to changing climatic conditions and mitigating risks requires a concerted effort that transcends political, economic, and cultural boundaries. By embracing strategies that bolster resilience, curb emissions, and foster sustainability, we pave the way toward a future that thrives in the face of uncertainty. As stewards of the planet, our choices today shape the legacy we leave for generations to come—a legacy of resilience, innovation, and a shared commitment to nurturing our planet's health.

Case studies of communities, industries, and regions implementing climate adaptation measures

Climate Adaptation in Action: Case Studies of Resilient Communities, Industries, and Regions

The global call to address climate change has spurred a wave of innovative adaptation measures, offering a glimpse into the possibilities of building a more climate-resilient world. From communities on the frontlines of rising sea levels to industries navigating changing landscapes, these case studies showcase the diverse ways in which adaptation strategies are being implemented to mitigate risks and ensure a sustainable future.

Case Study 1: The Netherlands' Flood-Ready Cities The Netherlands, renowned for its intricate system of dikes and polders, has embraced a proactive approach to safeguarding its cities from rising sea levels. As a low-lying country, it faces the constant threat of flooding. The Room for the River project, launched after the devastating 1995 floods, exemplifies the country's commitment to adaptation. By creating more space for rivers to flow during high-water periods, the project reduces flood risks and enhances urban resilience.

Case Study 2: Australian Great Barrier Reef Resilience Australia's Great Barrier Reef, an ecological marvel, faces the dual threats of warming waters and coral bleaching. The Reef 2050 Plan, a collaboration between government, industry, and community stakeholders, focuses on enhancing the Reef's resilience.

Strategies include reducing pollution runoff, managing coastal development, and implementing innovative coral restoration techniques to foster the Reef's ability to adapt to changing conditions.

Case Study 3: Toyota's Sustainable Mobility Initiative Industries are also stepping up to the challenge of climate adaptation. Toyota's Sustainable Mobility initiative aligns with global climate goals by emphasizing electric and hybrid vehicles. The company has committed to producing nearly ten million electrified vehicles by 2030, significantly reducing carbon emissions from the transportation sector. This forward-looking approach not only mitigates climate risks but also positions Toyota as a leader in the evolving automotive landscape.

Case Study 4: Bhutan's Carbon-Negative Commitment Bhutan, a small Himalayan nation, has made a remarkable commitment to being carbon-negative by absorbing more carbon dioxide than it emits. The country's unique focus on forest conservation and sustainable development contributes to its carbon-negative status. By prioritizing renewable energy sources and maintaining its lush forests, Bhutan showcases the potential for countries to take bold climate action.

Case Study 5: The Rebirth of Rotterdam's Floating Communities In response to sea level rise, the city of Rotterdam in the Netherlands is embracing an innovative approach: floating architecture. The Water Square project integrates sustainable urban planning with flood adaptation. Plazas designed to collect and store excess rainwater during heavy rainfall events help prevent flooding while enhancing public spaces. These adaptive measures underscore the city's commitment to both function and aesthetics in the face of changing climates.

Case Study 6: California's Drought-Resilient Water Management California's history of droughts has prompted the state to adopt comprehensive water management strategies. The Sustainable

Groundwater Management Act aims to stabilize declining groundwater levels by implementing groundwater recharge and sustainable use practices. By promoting water efficiency, reusing wastewater, and bolstering water storage, California is forging a path toward a more drought-resilient future.

Case Study 7: Bangladesh's Cyclone Preparedness Bangladesh, a country susceptible to cyclones and flooding, has implemented a community-based cyclone preparedness program. The Cyclone Preparedness Program (CPP) trains volunteers to disseminate early warnings, conduct evacuations, and provide emergency assistance during cyclone events. This grassroots approach has significantly reduced the loss of life during cyclones and showcases the power of community-driven adaptation efforts.

Shaping Our Destiny: Adaptation as a Shared Endeavor These case studies illuminate the diverse ways in which adaptation measures are being implemented to confront the challenges of a changing climate. From international agreements to localized community initiatives, the commitment to resilience spans scales and sectors. By learning from these examples and fostering a spirit of collaboration, we embrace the potential to navigate a future of uncertainty with ingenuity, determination, and a shared commitment to safeguarding our planet's well-being.

The importance of resilient infrastructure, sustainable agriculture, and disaster preparedness

Building a Climate-Ready World: The Pillars of Resilient Infrastructure, Sustainable Agriculture, and Disaster Preparedness

In an era of shifting climates and unpredictable weather patterns, the concepts of resilient infrastructure, sustainable agriculture, and disaster preparedness have emerged as critical cornerstones of a sustainable and adaptable future. As we confront the challenges of a changing world, these pillars offer not only protection against the impacts of climate change but also pathways toward thriving communities, food security, and enhanced resilience.

Resilient Infrastructure: Fortifying Against the Unpredictable
Resilient infrastructure encompasses the planning, design, and construction of systems that can withstand the shocks and stresses of a changing climate. From roads and bridges to energy grids and water supply networks, the durability of our built environment is paramount. By factoring in climate projections and extreme weather scenarios, we can ensure that our infrastructure remains operational during floods, hurricanes, and heatwaves. Resilient infrastructure not only safeguards lives and livelihoods but also reduces recovery costs and disruption to society.

Sustainable Agriculture: Nourishing a Growing Planet

Sustainable agriculture is the linchpin of global food security, and its role becomes even more vital as climate conditions shift. Adaptive agricultural practices prioritize soil health, water conservation, and biodiversity. Diverse cropping systems, precision agriculture, and agroforestry help farmers navigate changing weather patterns and evolving growing conditions. By fostering resilience in food production, sustainable agriculture ensures that communities have access to nutritious and reliable food sources, even in the face of climate-related challenges.

Disaster Preparedness: Anticipating the Unforeseen Disasters—both natural and human-induced—can have devastating impacts on communities. Disaster preparedness involves proactive planning, early warning systems, and coordinated responses to mitigate these impacts. Effective disaster preparedness not only saves lives but also reduces economic losses and social disruption. Investing in disaster-resistant infrastructure, conducting regular drills, and ensuring that communities have access to emergency resources are integral components of a resilient society.

Synergy and Interdependence: A Holistic Approach These three pillars—resilient infrastructure, sustainable agriculture, and disaster preparedness—are not isolated strategies; they are interconnected and mutually reinforcing. For instance, resilient infrastructure supports the transportation and distribution networks necessary for sustainable agriculture. Sustainable agricultural practices can contribute to disaster risk reduction by stabilizing landscapes and reducing soil erosion. Disaster preparedness efforts are more effective when supported by resilient infrastructure and sustainable food systems.

An Investment in Our Future: Long-Term Benefits Embracing these pillars is an investment in our planet's future, a commitment to nurturing a world that thrives in the face of uncertainty. Resilient infrastructure, sustainable agriculture, and disaster preparedness contribute to economic stability, community well-being, and environmental stewardship. They empower us to adapt to changing conditions, reduce vulnerabilities, and build a sustainable legacy for generations to come.

A Blueprint for Tomorrow: Guided by Wisdom As we navigate the complexities of climate change, these pillars offer a blueprint for action—one rooted in science, innovation, and a deep understanding of the interconnectedness of our world. By integrating resilience into our infrastructure, cultivating sustainable food systems, and preparing for disasters, we honor the delicate balance of our planet and reaffirm our commitment to shaping a future of vitality and prosperity.

The role of climate models in simulating past, present, and future climate scenarios

Modeling Earth's Climate: Simulating Past, Present, and Future Scenarios

In the intricate dance of Earth's climate system, understanding its dynamics is a monumental challenge. Enter climate models—a scientific tool that harnesses the power of mathematics, physics, and computer science to simulate the complex interactions that shape our planet's climate. These models serve as time machines, enabling us to explore the past, interpret the present, and anticipate potential futures.

Unraveling the Past: Simulating Paleoclimates Climate models offer a virtual window into the Earth's past, allowing us to simulate climates of epochs long gone. By inputting data from ice cores, sediment records, and other paleoclimatic proxies, these models can recreate conditions from thousands to millions of years ago. They unveil ancient temperature patterns, atmospheric compositions, and ocean circulation dynamics. By comparing model outputs with proxy data, scientists validate the models' accuracy and refine our understanding of Earth's climatic history.

Decoding the Present: Real-Time Insights In the here and now, climate models serve as invaluable tools for interpreting and predicting current climate conditions. By assimilating vast datasets—ranging from atmospheric measurements to oceanic temperatures—models generate real-time simulations of Earth's climate. These simulations inform weather forecasts, track the progression of phenomena like El Niño and La Niña, and provide a comprehensive view of today's climate dynamics. The accuracy of these predictions continuously improves as models are refined with new data and advanced methodologies.

Forecasting the Future: Peering Ahead The true power of climate models lies in their ability to project into the future. By considering factors such as greenhouse gas emissions, land use changes, and solar radiation variations, models generate scenarios that offer insights into potential climate trajectories. These projections inform our understanding of how global temperatures might rise, sea levels could change, and precipitation patterns may shift in the coming decades and centuries. They guide policymakers, researchers, and societies in developing strategies to adapt to and mitigate the impacts of climate change.

The Dance of Variables: Building Complex Models Climate models simulate the interplay of myriad variables—temperature, humidity, atmospheric pressure, ocean currents, and more. These models are built upon the foundation of physical laws and principles governing Earth's systems. They divide the planet into three-dimensional grids, applying mathematical equations to each grid cell to calculate how variables evolve over time. Supercomputers crunch these calculations, producing intricate and dynamic simulations of Earth's climate behavior.

Navigating Uncertainty: The Challenge of Complex Systems Climate models are not crystal balls; they are tools that approximate the intricacies of Earth's climate. Uncertainties arise from various sources, including the inherent complexity of the climate system, the limitations of available data, and the potential influence of unknown factors. However, models are designed to encompass a range of scenarios, allowing scientists to assess potential outcomes under different conditions and assumptions.

A Roadmap for Action: Informed Decision-Making Climate models empower us to make informed decisions in the face of a changing world. They guide policies aimed at mitigating greenhouse gas emissions, preparing for sea level rise, and safeguarding communities against extreme weather events. These models are not just simulations; they are blueprints for a sustainable and resilient future. Through the interplay of science, technology, and human ingenuity, we leverage the power of modeling to shape a world that navigates climate challenges with wisdom and foresight.

Techniques for predicting climate patterns, trends, and potential impacts

Predicting Climate Patterns and Trends: Charting the Course Ahead

As humanity grapples with the complexities of a changing climate, the ability to predict climate patterns, trends, and potential impacts has become essential for informed decision-making and effective planning. Employing an array of advanced techniques, scientists are uncovering valuable insights into Earth's future climate trajectories, allowing us to anticipate challenges, adapt to shifts, and mitigate risks.

Statistical Analyses: Mining Historical Data Statistical methods analyze historical climate data to identify patterns and trends. Time-series analyses, for instance, reveal cyclical behaviors like seasonal changes and long-term variations. These techniques allow scientists to project trends forward, providing a sense of how temperatures, precipitation patterns, and other climate variables might evolve.

Climate Models: Simulating Complex Dynamics Climate models are virtual laboratories that simulate the interactions of Earth's systems. These sophisticated computer models integrate data on atmospheric, oceanic, and land processes to generate projections of future climate scenarios. By varying input parameters like greenhouse gas emissions and solar radiation, models offer a range of potential outcomes, helping us understand possible climate trajectories.

Paleoclimatic Proxies: Gleaning Insights from the Past The Earth's history holds clues to its future, and paleoclimatic proxies are the key to unlocking these insights. Ice cores, tree rings, sediment layers, and other proxies provide records of past climate conditions. By studying these records, scientists uncover patterns of natural variability and use them as a basis for projecting future climate trends.

Ensemble Approaches: Weighing Multiple Scenarios Predicting climate patterns demands an acknowledgment of uncertainty. Ensemble approaches involve running multiple climate models with slight variations in input parameters. By examining the range of outcomes generated by these models, scientists can quantify the level of uncertainty associated with different projections and identify trends that are more likely to occur.

Emerging Technologies: Harnessing Big Data and AI The era of big data and artificial intelligence is transforming climate prediction. Machine learning algorithms analyze vast datasets, identifying hidden patterns and relationships. These techniques enhance the accuracy of predictions by uncovering complex interactions that might be overlooked using traditional methods.

Global Observations: Real-Time Monitoring A comprehensive understanding of climate patterns requires continuous global observations. Networks of weather stations, satellites, buoys, and sensors provide real-time data on temperature, precipitation, ocean currents, and more. These observations allow scientists to monitor ongoing changes, validate models, and improve prediction accuracy.

Impact Assessments: Connecting Climate to Consequences Predicting climate patterns is only part of the puzzle. Understanding the potential impacts on ecosystems, agriculture, water resources, and communities is equally crucial. Integrated assessment models combine climate projections with socioeconomic data to project how climate change might affect various sectors, helping policymakers make informed decisions.

A Path Forward: Informed Action The ability to predict climate patterns and trends empowers us to take proactive steps toward a sustainable future. From adapting infrastructure to planning for shifts in agricultural zones, these predictions enable us to navigate the challenges and opportunities presented by a changing climate. By integrating science, technology, and human ingenuity, we leverage prediction techniques to make decisions that safeguard our planet's health and well-being.

Challenges and uncertainties in climate modeling and the importance of improving accuracy

Navigating the Complexities: Challenges and Uncertainties in Climate Modeling

Climate modeling is a powerful tool that offers valuable insights into Earth's complex climate systems. However, as with any endeavor involving intricate interactions and vast data, challenges and uncertainties abound. Recognizing and addressing these limitations is vital for improving the accuracy of climate models and enhancing their role in shaping our understanding of climate change.

Incomplete Data: Gaps in Understanding Climate models rely on a wealth of data from various sources, including historical records, satellite observations, and paleoclimatic proxies. However, gaps in data collection, inconsistencies in measurements, and limited historical records can hinder model accuracy. In regions with sparse monitoring networks, accurately capturing local climate dynamics becomes challenging.

Complex Interactions: Nature's Complexity Earth's climate is a complex web of interrelated processes, involving the atmosphere, oceans, land surfaces, and more. Modeling the intricate interactions among these components is a daunting task, as even small variations in one system can have cascading effects on others. The challenge lies in accurately representing these interactions in a mathematical framework.

Uncertain Feedbacks: Amplifying Effects Climate models must consider various feedback mechanisms, where changes in one variable trigger responses that amplify or dampen the original change. For instance, as temperatures rise, melting ice reduces surface reflectivity, leading to further warming. Predicting the extent and magnitude of these feedbacks introduces uncertainties that can affect long-term projections.

Model Resolution: Bridging the Gap The resolution of climate models—the granularity with which they simulate Earth's systems—poses challenges. Global climate models divide the planet into grid cells, and finer resolution requires immense computational power. Balancing the need for finer detail with computational feasibility is a delicate task, particularly when modeling regional or localized impacts.

Complex Processes: Atmospheric Aerosols and Clouds The interactions of atmospheric aerosols and clouds are among the most challenging aspects of climate modeling. Aerosols can either cool or warm the atmosphere, and clouds play a role in reflecting sunlight and trapping heat. These processes are influenced by a myriad of factors, including aerosol composition, cloud formation mechanisms, and regional air quality.

Natural Variability: Distinguishing Trends Climate models must distinguish between natural variability and human-induced changes. Variability driven by phenomena like El Niño can mask or exaggerate long-term trends. Isolating and attributing observed changes to specific causes is a complex task that requires meticulous analysis and validation against historical data.

Future Scenarios: Predicting Human Behavior Predicting future climate scenarios requires assumptions about future greenhouse gas emissions, technological advancements, and societal behaviors. These assumptions introduce uncertainties, as they depend on human choices that are influenced by a wide range of factors, including policy decisions, economic developments, and social attitudes.

The Pursuit of Improvement: Enhancing Accuracy Despite these challenges, the importance of improving the accuracy of climate models cannot be overstated. Enhanced accuracy enables more informed decision-making, more effective climate policies, and a better understanding of potential impacts. Addressing uncertainties involves refining model physics, expanding observational networks, validating models against historical data, and incorporating new scientific findings.

A Shared Endeavor: Science, Technology, and Collaboration The complexity of climate modeling underscores the need for collaboration among scientists, technologists, policymakers, and communities. As models become more sophisticated and computational capabilities expand, collaboration becomes pivotal in refining models, reducing uncertainties, and harnessing the power of modeling to shape a more resilient and sustainable future.

Exploration of the historical and cultural significance of climate in different societies

Climate Through the Lens of History and Culture: A Tapestry of Significance

The climate has always been more than just a backdrop to human history and culture—it has woven its intricate threads into the very fabric of societies around the world. Across civilizations and epochs, climate has influenced beliefs, shaped economies, driven migrations, and shaped the narratives that define who we are. Exploring its historical and cultural significance unveils a tapestry of diverse perspectives, lessons, and connections.

Ancient Civilizations: Navigating the Elements From the banks of the Nile to the valleys of the Indus, ancient civilizations were deeply attuned to the rhythms of climate. The Nile's annual floods, driven by monsoons and equatorial rains, were vital to sustaining ancient Egypt's agricultural bounty. Similarly, the cycles of the monsoon season played a central role in shaping the cultural practices of societies in South Asia, including festivals, rituals, and trade.

Nomadic Cultures: Following the Seasons For nomadic cultures, climate dictated their way of life. The migratory patterns of pastoral communities were intricately tied to the availability of grazing lands and water sources. As seasons shifted, so did their paths, leading to a dynamic relationship with the land and its resources. This constant movement influenced their traditions, folklore, and survival strategies.

Cultural Symbols: Climate in Art and Myth Climate often found its way into art, literature, and mythologies. In ancient Greece, the changing seasons were reflected in the story of Demeter and Persephone, while Native American cultures wove tales of the Thunderbird controlling rains and storms. The powerful symbolism of climate in these narratives offered explanations for natural phenomena and underscored the profound connection between humans and the environment.

Economic Dynamics: Weathering the Storms Throughout history, climatic variations have triggered economic shifts. The "Little Ice Age" that spanned the late Middle Ages to the early modern period influenced agricultural productivity, altered trade routes, and even contributed to political upheavals. The Dutch Golden Age, for example, was shaped by the favorable climatic conditions that allowed for increased agricultural output and trade.

Migration and Settlement: Shaping the Landscapes Climate fluctuations have often forced human migrations and reshaped settlement patterns. The desertification of the Sahel region in Africa, for instance, led to the movement of communities and the emergence of new cultural practices. The Ancestral Puebloans' migration from the Mesa Verde region in the American Southwest is another example, illustrating how climate played a role in shaping the architecture and lifestyles of these societies.

Traditional Knowledge: Wisdom of the Elders Indigenous communities hold invaluable traditional knowledge about climate and the environment. Passed down through generations, this knowledge encompasses weather prediction, resource management, and adaptation strategies. Indigenous practices demonstrate a deep understanding of local ecosystems and the interplay between nature and culture.

Contemporary Reflections: Climate's Societal Impact In the modern era, climate change has spurred conversations that transcend borders. Rising sea levels threaten cultural heritage sites, forcing societies to grapple with the preservation of their past. Indigenous communities are on the frontlines of climate change, advocating for sustainable practices rooted in their deep connections to the land.

From Lessons to Actions: A Global Effort Exploring the historical and cultural significance of climate unveils a shared heritage and a collective responsibility. By learning from the resilience, adaptability, and insights of diverse societies, we are better equipped to address the challenges of a changing climate. The lessons embedded in the narratives of the past echo into our present and guide our steps toward a more sustainable and harmonious future.

Societal impacts of climate change, including migration, conflicts, and economic shifts

Unraveling Societal Fabric: Climate Change's Far-Reaching Impact

Climate change is more than an environmental challenge —it's a multi-faceted phenomenon that reverberates through societies, altering the very foundations of human existence. As temperatures rise, sea levels creep higher, and weather patterns become more erratic, the intricate web of societal dynamics experiences profound shifts, including migration, conflicts, and economic transformations. Navigating these impacts requires a deep understanding of the complex interplay between the environment and human societies.

Climate-Induced Migration: Displacement and Resettlement
Rising sea levels, extreme weather events, and diminishing agricultural yields can render once-viable habitats uninhabitable. This forces communities to migrate in search of safer and more sustainable living conditions. Coastal residents are often among the first to face this reality, with island nations and low-lying coastal areas being particularly vulnerable. The displacement of people can strain resources in host regions and lead to challenges related to housing, employment, and integration.

Resource Scarcity and Conflicts: A Recipe for Tension As resources like water, arable land, and energy become scarcer due to climate change, competition for these essentials can escalate into conflicts. In regions already grappling with political instability or ethnic tensions, environmental stressors can exacerbate existing challenges. Conflicts over access to resources have the potential to fuel violence and displacement, amplifying the human toll of climate change.

Economic Shifts: Disruptions and Adaptations The impacts of climate change ripple through economies, altering sectors from agriculture and tourism to manufacturing and infrastructure. Extreme weather events can disrupt supply chains, damage infrastructure, and lead to economic losses. On the flip side, economic shifts driven by adaptation efforts—such as investments in renewable energy, climate-resilient agriculture, and green technologies—can create new opportunities and industries.

Health Implications: Vulnerabilities and Risks Climate change introduces health risks ranging from heat-related illnesses to the spread of diseases carried by vectors like mosquitoes. Heatwaves can exacerbate existing health conditions and strain healthcare systems. Additionally, changes in precipitation patterns can affect water quality and availability, leading to waterborne diseases. Vulnerable populations, particularly in low-income communities, are often hit hardest by these health impacts.

Cultural and Heritage Loss: Fading Identities As climate change erodes coastlines and intensifies storms, cultural heritage sites and indigenous knowledge can be threatened. Sea level rise can submerge historical sites, erasing tangible connections to the past. Indigenous communities, whose cultural practices and identities are intertwined with their lands, face the loss of ancestral territories due to environmental changes.

Gender and Social Inequalities: Amplifying Disparities Climate change can magnify existing gender and social inequalities. Women, who often bear the primary responsibility for household water and food security, are disproportionately affected by changing weather patterns. In some contexts, women may have limited access to decision-making processes related to climate adaptation and resource management.

Collective Responses: Addressing Complex Challenges The societal impacts of climate change necessitate collective responses that consider diverse contexts, cultures, and vulnerabilities. Mitigation efforts—such as reducing greenhouse gas emissions—and adaptation strategies—such as building climate-resilient infrastructure—are vital components of safeguarding societies. Strengthening social safety nets, investing in education, and promoting sustainable livelihoods can help communities withstand the disruptive forces of climate change.

A Shared Responsibility: Fostering Resilience As societies grapple with the profound shifts brought about by climate change, fostering resilience becomes paramount. This involves recognizing the interconnectedness of environmental, social, and economic systems and striving for solutions that prioritize the well-being of all. By acknowledging the far-reaching impacts of climate change and working collaboratively, we can create a more equitable and sustainable future for generations to come.

The role of education and awareness in fostering responsible climate stewardship

Empowering Change: Education and Awareness as Climate Stewards

In the face of the complex challenges posed by climate change, education and awareness emerge as powerful catalysts for responsible climate stewardship. The journey toward a sustainable future hinges on equipping individuals, communities, and societies with the knowledge, empathy, and motivation needed to effect positive change. By fostering a deeper understanding of the intricacies of our planet's climate systems, as well as the interconnectedness of human actions and environmental consequences, education and awareness play a pivotal role in shaping responsible climate stewardship.

Knowledge as Empowerment: Understanding Climate Science
Education is the cornerstone of empowerment. Equipping individuals with a fundamental grasp of climate science—covering topics such as greenhouse gas emissions, temperature trends, and the impacts of global warming—provides them with the tools to make informed decisions. Through formal education systems, workshops, and accessible resources, people can grasp the urgency of the climate crisis and comprehend the science behind it.

Personal Connection: Bridging the Gap Climate change can sometimes feel distant and overwhelming, but education helps bridge this gap by making it personal. Learning about the local and global impacts of climate change—such as rising sea levels, extreme weather events, and shifts in biodiversity—connects individuals to the broader narrative of environmental change. Personalizing these issues fosters empathy and encourages action.

Critical Thinking: Navigating Complex Choices Climate change is a multidimensional challenge that demands critical thinking. Education cultivates analytical skills that empower individuals to evaluate the validity of information, distinguish between fact and misinformation, and make well-informed decisions. In an era of misinformation, these skills are essential for steering society toward evidence-based solutions.

Behavioral Shifts: Translating Awareness into Action Awareness alone is not enough; translating knowledge into action is key. Education highlights the tangible ways individuals can reduce their carbon footprint—such as embracing renewable energy, adopting sustainable consumption practices, and reducing waste. By understanding how personal choices contribute to collective impacts, individuals can make conscious decisions that mitigate climate change.

Engagement and Advocacy: Amplifying Impact Educational initiatives also cultivate a sense of agency and encourage active participation. As people grasp the significance of climate change, they are more likely to engage in advocacy efforts, support policies that prioritize sustainability, and participate in community initiatives. Education fosters a sense of responsibility to safeguard the planet for future generations.

Interdisciplinary Insights: Collaborative Solutions Climate change is not confined to a single discipline—it encompasses science, economics, policy, and ethics. Education provides a platform for interdisciplinary learning, enabling individuals to approach climate issues from various angles and collaborate across fields. This holistic perspective is essential for crafting comprehensive solutions.

Cultural Awareness: Embracing Diversity Education and

awareness initiatives that are culturally sensitive and inclusive resonate with diverse audiences. By acknowledging cultural perspectives and traditions, these efforts encourage communities to integrate climate-conscious practices into their ways of life. This approach fosters a sense of ownership and facilitates the adoption of sustainable behaviors.

A Catalyst for Transformation: Collective Impact The role of education and awareness extends beyond individual empowerment; it fuels a collective movement toward responsible climate stewardship. As people become educated advocates, informed consumers, and engaged citizens, the momentum for change gains strength. By fostering a culture of sustainability, education and awareness create a ripple effect that extends from homes and classrooms to policy halls and global stages.

Guardians of Tomorrow: Nurturing Climate Champions Education and awareness are potent tools for nurturing a generation of climate champions—individuals who understand the urgency of the climate crisis, are equipped with the knowledge to address it, and are committed to fostering a harmonious relationship between humanity and the planet. As we navigate the uncharted waters of climate change, education and awareness serve as guiding lights, illuminating a path toward a more sustainable and resilient world.

The imperative for global cooperation in addressing climate change

United for Change: The Imperative of Global Cooperation in Addressing Climate Change

Climate change is a global challenge that transcends borders, economies, and ideologies. Its far-reaching impacts remind us that the fate of our planet is interconnected and interdependent. In the face of such a monumental crisis, the imperative for global cooperation becomes clear—only by working together can humanity effectively mitigate and adapt to the challenges posed by climate change. Collaboration on a global scale is not just an option; it is the only viable path forward.

Shared Responsibility: One Planet, One Goal Climate change does not discriminate between nations. Rising temperatures, sea level rise, and extreme weather events affect all corners of the world, regardless of national boundaries. As inhabitants of a shared planet, we bear a collective responsibility to safeguard its health for present and future generations. This shared responsibility forms the foundation for international cooperation.

Collective Impact: Leveraging Diverse Expertise Global cooperation harnesses the power of diverse expertise and resources. Each nation brings unique strengths—technological innovations, policy insights, financial resources—to the table. Collaborating across borders allows us to pool knowledge and skills, leading to more effective solutions and faster progress.

Transcending Political Divides: A Common Cause Climate

change offers an opportunity to transcend political divides and forge common ground. It is a crisis that affects every nation, irrespective of political systems or ideologies. By focusing on shared goals, countries can set aside differences and work collectively toward a sustainable and resilient future.

Mitigation and Adaptation: Bridging the Gap Global cooperation is vital for both mitigation and adaptation efforts. Mitigation—reducing greenhouse gas emissions—requires coordinated efforts to transition to cleaner energy sources, implement sustainable practices, and develop innovative technologies. Adaptation—building resilience to the impacts already underway—demands knowledge-sharing, support for vulnerable communities, and collaborative efforts to develop climate-resilient infrastructure.

International Commitments: Setting the Stage International agreements, such as the Paris Agreement, serve as critical frameworks for global climate action. These agreements establish common goals, encourage transparency, and create a platform for countries to hold each other accountable. They provide a roadmap for cooperation, ensuring that no nation is left behind in the pursuit of a sustainable future.

Transboundary Challenges: Recognizing Interconnectedness Certain climate impacts transcend national borders. Sea level rise, for instance, affects coastal regions worldwide, regardless of the contributing nations. Addressing these transboundary challenges requires cooperation to develop strategies that protect vulnerable communities and ecosystems.

Climate Justice: Addressing Inequities Global cooperation is essential for addressing climate justice—ensuring that the burdens and benefits of climate action are distributed fairly. Developing nations, often the least responsible for emissions, are disproportionately affected by climate impacts. Collaborative efforts can provide support for adaptation, technology transfer, and capacity-building, ensuring that vulnerable communities are not left to face the crisis alone.

A Call to Action: Leaving a Legacy of Unity The urgent need for global cooperation in addressing climate change is a call to action for nations, organizations, and individuals alike. It is a call to transcend short-term interests and work toward a sustainable legacy of unity, resilience, and responsible stewardship. By recognizing our shared fate and pooling our resources, knowledge, and determination, we can forge a brighter future for all, leaving a legacy of solidarity and collaboration in the face of one of humanity's greatest challenges.

Policies, agreements, and initiatives aimed at reducing greenhouse gas emissions

Charting a Greener Path: Policies, Agreements, and Initiatives to Reduce Greenhouse Gas Emissions

As the threat of climate change becomes increasingly urgent, governments, organizations, and communities around the world are taking proactive steps to curb greenhouse gas emissions and transition to a more sustainable future. Through a combination of policies, international agreements, and collaborative initiatives, these efforts are shaping a greener and more resilient world.

The Paris Agreement: A Global Commitment One of the most significant milestones in international climate action is the Paris Agreement. Adopted in 2015 under the United Nations Framework Convention on Climate Change (UNFCCC), the agreement brings together nearly every country on Earth in a common effort to combat climate change. Signatories commit to limiting global warming to well below 2 degrees Celsius above pre-industrial levels, with efforts to limit it to 1.5 degrees—a threshold critical for avoiding the most severe impacts of climate change. Countries pledge to regularly update their targets for reducing emissions and provide financial support to developing nations for both mitigation and adaptation efforts.

Renewable Energy Targets: A Shift in Power Many nations are setting ambitious targets for renewable energy deployment. These targets, often backed by policies such as feed-in tariffs, tax incentives, and renewable portfolio standards, aim to transition away from fossil fuels and toward cleaner energy sources like solar, wind, and hydroelectric power. Countries are investing in renewable energy infrastructure, advancing technology, and promoting energy efficiency to accelerate the shift to low-carbon energy systems.

Carbon Pricing: Incentivizing Emission Reductions Carbon pricing mechanisms put a price on carbon emissions, creating economic incentives for industries to reduce their greenhouse gas output. These mechanisms can take the form of carbon taxes or cap-and-trade systems. Carbon pricing not only encourages emission reductions but also generates revenue that can be reinvested in renewable energy projects and climate adaptation efforts.

Electric Mobility Initiatives: Driving Change The transportation sector is a significant contributor to greenhouse gas emissions. Initiatives promoting electric vehicles (EVs) and charging infrastructure are gaining traction worldwide. Governments are offering incentives for EV adoption, including tax credits, rebates, and reduced registration fees. Additionally, some regions are investing in public transportation networks, cycling infrastructure, and pedestrian-friendly urban planning to reduce reliance on fossil fuel-powered vehicles.

Reforestation and Afforestation Programs: Breathing Life into Landscapes Trees and forests play a vital role in sequestering carbon dioxide from the atmosphere. Reforestation and afforestation initiatives focus on planting and restoring forests to enhance carbon sinks. These efforts not only mitigate climate change but also provide additional ecological benefits, such as biodiversity conservation and natural resource protection.

International Climate Initiatives: Collaborative Action Beyond the Paris Agreement, various international initiatives foster collaborative action. The "Mission Innovation" initiative aims to accelerate clean energy innovation and investment among participating nations. The "Global Alliance for Buildings and

Construction" focuses on making the building sector more energy-efficient. These initiatives demonstrate how international collaboration can drive meaningful change across sectors.

Local and Regional Initiatives: Leading from the Ground Up Cities, states, and provinces are taking bold steps to reduce emissions and increase climate resilience. Many have adopted climate action plans that include targets for renewable energy adoption, energy efficiency improvements, and emissions reductions. Local initiatives can lead to innovative solutions tailored to specific regional challenges.

Corporate Sustainability Pledges: Business for Change Many companies recognize the importance of addressing climate change and are making commitments to reduce their emissions. Corporate sustainability pledges often involve setting emissions reduction targets, transitioning to renewable energy sources, and adopting sustainable practices across supply chains.

A Collective Effort: Shaping a Sustainable Future Policies, agreements, and initiatives aimed at reducing greenhouse gas emissions represent a collective effort to address the pressing challenges of climate change. As governments, organizations, and individuals work together, these actions contribute to the broader mission of safeguarding our planet and ensuring a more sustainable and resilient future for generations to come.

Individual and collective actions to mitigate climate change's effects and ensure a sustainable future

Empowering Change: Individual and Collective Actions for a Sustainable Future

The urgency of mitigating climate change's effects has sparked a global movement of individuals, communities, and organizations committed to creating a more sustainable world. From small everyday choices to collective initiatives, these actions are driving transformative change, highlighting the power of collective efforts to ensure a sustainable future for our planet.

Individual Actions: Making Every Choice Count

1. **Energy Efficiency:** Reducing energy consumption at home by using energy-efficient appliances, LED lighting, and proper insulation.
2. **Transportation Choices:** Opting for public transportation, carpooling, biking, or walking instead of driving alone, and considering electric or fuel-efficient vehicles.
3. **Reducing Waste:** Embracing the "reduce, reuse, recycle" mantra by minimizing single-use plastics, composting organic waste, and recycling properly.
4. **Sustainable Diet:** Choosing plant-based and locally sourced foods to reduce the environmental impact of diet choices.
5. **Water Conservation:** Conserving water by fixing leaks,

using water-saving fixtures, and practicing mindful water use.

6. **Eco-Friendly Purchases:** Supporting sustainable products and brands that prioritize ethical sourcing, minimal packaging, and fair labor practices.
7. **Energy Sources:** Advocating for and switching to renewable energy sources for home and business needs.
8. **Climate Education:** Staying informed about climate science and sharing knowledge with friends, family, and online communities.

Community and Collective Actions: Amplifying Impact

1. **Local Advocacy:** Joining or supporting local environmental organizations and community groups that work on climate-related issues.
2. **Green Infrastructure:** Advocating for and participating in tree planting, urban gardening, and green space development in urban areas.
3. **Climate-Friendly Policies:** Engaging with local and national governments to support policies that promote renewable energy, sustainable transportation, and emissions reduction.
4. **Sustainable Businesses:** Supporting businesses that prioritize sustainability, whether by purchasing from them or encouraging others to do so.
5. **Educational Campaigns:** Participating in and organizing events, workshops, and campaigns to raise awareness about climate change and sustainable living.
6. **Youth Initiatives:** Encouraging and supporting youth-led movements and initiatives focused on climate action.
7. **Collective Impact:** Joining international initiatives, such as "Earth Hour" or "Climate Week," to showcase collective commitment to climate action.
8. **Collaboration and Partnerships:** Building partnerships

with diverse stakeholders, including businesses, NGOs, governments, and communities, to tackle climate challenges together.

Corporate and Organizational Actions: Leading by Example

1. **Carbon Neutrality:** Setting ambitious goals to achieve carbon neutrality or net-zero emissions by a certain date.
2. **Renewable Energy Adoption:** Investing in renewable energy sources to power operations and supply chains.
3. **Circular Economy Practices:** Implementing circular economy models that reduce waste, reuse materials, and promote sustainable product lifecycles.
4. **Eco-Friendly Design:** Incorporating sustainable design principles into products, buildings, and infrastructure projects.
5. **Sustainable Supply Chains:** Partnering with suppliers that adhere to ethical and sustainable practices.
6. **Environmental Reporting:** Transparently disclosing environmental impacts, emissions data, and sustainability efforts.
7. **Employee Engagement:** Encouraging and supporting employees in adopting sustainable practices both at work and in their personal lives.
8. **Innovation and Research:** Investing in research and development to create innovative technologies and solutions that address climate challenges.

Global Cooperation: Navigating a Collective Path

1. **International Agreements:** Advocating for strong international agreements and commitments to address climate change at the global level.
2. **Climate Finance:** Supporting initiatives that provide financial resources to developing countries for climate adaptation and mitigation efforts.

3. **Technology Transfer:** Sharing renewable energy and sustainability technologies with developing nations to accelerate global progress.

4. **Climate Diplomacy:** Participating in diplomatic efforts to foster collaboration, information sharing, and policy alignment among nations.

A Unified Force for Change: Together Toward Sustainability
Whether as individuals, communities, or organizations, the actions taken to mitigate climate change's effects are a testament to the power of collective commitment. By fostering a culture of responsibility, innovation, and collaboration, we are not just responding to a challenge—we are building a foundation for a sustainable future where the health of our planet and the well-being of all its inhabitants are paramount.

Advances in climate science and technology, including remote sensing and data analytics

Advances in Climate Science and Technology: Pioneering a Sustainable Future

In the quest to understand, predict, and mitigate the impacts of climate change, remarkable advances in both climate science and technology are shaping our ability to tackle this complex global challenge. Cutting-edge tools such as remote sensing and data analytics are revolutionizing our approach to climate research, enabling us to gather crucial insights, make informed decisions, and pave the way toward a more sustainable future.

Remote Sensing: A Window to the Earth

1. **Satellite Observations:** Satellites equipped with advanced sensors provide a bird's-eye view of our planet's climate system. They track atmospheric composition, sea surface temperatures, ice cover, deforestation, and urban expansion, generating a wealth of data that informs climate models and policy decisions.

2. **Climate Monitoring:** Remote sensing enables real-time monitoring of key climate indicators, such as greenhouse gas concentrations, ocean temperature trends, and ice melt rates. This data informs both short-term weather predictions and long-term climate projections.

3. **Deforestation and Land Use:** Remote sensing aids in tracking deforestation rates and changes in land

use, crucial for understanding the carbon balance of ecosystems and addressing biodiversity loss.

4. **Early Warning Systems:** By identifying patterns in temperature, precipitation, and sea level rise, remote sensing can contribute to the development of early warning systems for extreme weather events, allowing communities to prepare and respond more effectively.

Data Analytics: Unleashing Insights from Big Data

1. **Climate Modeling:** Data analytics techniques process vast amounts of climate data to refine and improve climate models. This aids in simulating various scenarios, projecting future climate trends, and assessing the impacts of policy decisions.
2. **Extreme Event Attribution:** Advanced data analytics can attribute specific weather events, such as heatwaves or hurricanes, to climate change, helping us understand the link between human activity and extreme weather.
3. **Predictive Analysis:** Combining historical climate data with machine learning algorithms enhances our ability to predict shifts in climate patterns, which is essential for adapting to changing conditions.
4. **Risk Assessment:** Data analytics plays a key role in assessing climate-related risks, whether for industries, communities, or ecosystems. This enables more informed decision-making and targeted risk reduction strategies.

Climate-Technology Synergy: Paving the Way Forward

1. **Climate Finance:** Technological advancements facilitate accurate measurement and verification of emissions reductions, critical for carbon trading and incentivizing climate-friendly investments.
2. **Climate Adaptation:** Climate models, powered by technology, provide insights into regional

vulnerabilities and adaptation strategies. This information guides infrastructure planning, water management, and disaster preparedness.

3. **Policy Formulation:** Data-driven insights enable policymakers to design more effective climate policies and regulations. Decision-makers can assess the potential impacts of policy options and identify areas of intervention.

4. **Public Awareness:** Technology enhances communication of complex climate science to the public. Interactive visualizations, online platforms, and apps engage people in understanding the science behind climate change and motivate sustainable actions.

Challenges and Ethical Considerations: A Balanced Approach
While advances in climate science and technology offer tremendous opportunities, they also raise important challenges and ethical considerations:

1. **Data Privacy:** Collecting and sharing climate data must adhere to strict ethical standards to protect individual privacy.

2. **Data Bias:** Ensuring the accuracy and inclusivity of data is crucial to prevent biases in analyses that could disproportionately affect vulnerable populations.

3. **Equity:** Climate science and technology should be accessible to all nations and communities, regardless of their economic status.

4. **Ethical AI:** The use of artificial intelligence in climate research requires careful consideration to ensure fairness and transparency in decision-making.

Toward a Sustainable Tomorrow: Harnessing Innovation for Change The convergence of climate science and technology has ushered in a new era of possibility in our battle against climate change. By leveraging the insights gained from remote sensing and data analytics, we can design more effective mitigation and adaptation strategies, promote sustainable behaviors, and empower individuals, organizations, and governments to work together toward a sustainable future.

Innovative solutions for sustainable energy, carbon capture, and climate adaptation

Innovative Solutions for a Sustainable Future: Energy, Carbon Capture, and Climate Adaptation

As the challenges posed by climate change intensify, innovation becomes a driving force behind developing sustainable solutions. From harnessing clean energy sources to implementing carbon capture technologies and adapting to the changing climate, groundbreaking innovations are shaping a more resilient and environmentally conscious world.

Sustainable Energy Solutions: Powering the Future Responsibly

1. **Renewable Energy Integration:** The integration of solar, wind, hydro, and geothermal energy into the grid is revolutionizing the energy landscape, reducing reliance on fossil fuels and lowering greenhouse gas emissions.
2. **Energy Storage:** Advancements in battery technology are making energy storage more efficient and cost-effective. Energy storage solutions help balance supply and demand, ensuring a steady flow of renewable energy.
3. **Smart Grids:** Smart grids use data analytics and automation to optimize energy distribution, reduce waste, and enhance grid reliability, allowing for better integration of renewable energy sources.
4. **Energy Efficiency Innovations:** Innovative building materials, energy-efficient appliances, and smart

technologies are minimizing energy consumption in homes, commercial buildings, and industries.

Carbon Capture and Storage (CCS): Reducing Carbon Footprints

1. **Direct Air Capture:** Technologies that capture carbon dioxide directly from the air are being developed, offering the potential to remove CO_2 from the atmosphere and mitigate the effects of historical emissions.
2. **Enhanced Carbon Mineralization:** Certain minerals can react with carbon dioxide to form stable compounds, effectively sequestering CO_2. Innovations in this area aim to accelerate this natural process.
3. **Carbon Capture from Industrial Processes:** Innovations in capturing carbon emissions from industrial facilities, such as cement and steel production, play a crucial role in reducing industrial carbon footprints.
4. **Carbon Utilization:** Developing products from captured carbon, such as building materials or fuels, can provide economic incentives for carbon capture while reducing emissions.

Climate Adaptation Innovations: Building Resilience

1. **Nature-Based Solutions:** Implementing nature-based solutions, such as restoring wetlands, planting trees, and creating green infrastructure, helps communities adapt to changing climate conditions while enhancing biodiversity.
2. **Desalination Technologies:** Water scarcity is a growing concern in the face of climate change. Advanced desalination technologies provide a solution by converting seawater into freshwater for consumption and agriculture.
3. **Floating Architecture:** In flood-prone areas, floating structures and buildings designed to adapt to rising sea

levels are providing resilient housing solutions.

4. **Climate-Resilient Agriculture:** Crop varieties that are more tolerant to heat, drought, and changing climate conditions are being developed to ensure food security in a changing world.

5. **Early Warning Systems:** Innovative technology, including real-time weather monitoring and predictive analytics, assists in developing early warning systems for extreme weather events.

Innovative Financing and Collaborative Approaches: Paving the Way

1. **Green Bonds and Sustainable Investments:** Financial mechanisms like green bonds mobilize funds for climate-related projects, fostering investments in renewable energy, sustainable infrastructure, and climate resilience.

2. **Public-Private Partnerships:** Collaborations between governments, businesses, and nonprofits drive innovation by combining resources, expertise, and influence to accelerate climate solutions.

3. **Innovation Hubs and Accelerators:** Innovation hubs and accelerators provide a platform for startups and entrepreneurs to develop and scale climate-related technologies and solutions.

Barriers and the Way Forward: Overcoming Challenges

1. **Investment Challenges:** Scaling up innovative solutions requires substantial financial investment. Governments, private sectors, and international organizations must collaborate to provide funding.

2. **Regulatory Hurdles:** Supportive policies and regulatory frameworks are essential to incentivize the adoption of new technologies and the transition to cleaner energy sources.

3. **Technology Transfer:** Ensuring equitable access to innovative solutions across different regions requires technology transfer and knowledge sharing.
4. **Behavioral Change:** Overcoming resistance to change and promoting sustainable behaviors among individuals, communities, and industries is a significant challenge.

Catalyzing Transformation: An Opportunity for Progress
Innovative solutions in sustainable energy, carbon capture, and climate adaptation are driving transformation across sectors. The convergence of scientific discovery, technological innovation, and collective action offers a powerful path toward a more sustainable, resilient, and harmonious relationship between humanity and the planet we call home.

The potential for science and innovation to drive positive change and shape policy

Harnessing Science and Innovation: A Catalyst for Positive Change and Policy Evolution

In a rapidly changing world marked by pressing challenges such as climate change, resource depletion, and societal inequality, the role of science and innovation has become pivotal in driving positive change and shaping effective policy responses. Through rigorous research, groundbreaking technologies, and informed policymaking, science and innovation have the potential to reshape our world, address complex issues, and propel us toward a more sustainable and equitable future.

Informing Evidence-Based Policy:

1. **Rigorous Research:** Scientific research generates data-driven insights that inform policymakers about the intricacies of challenges and potential solutions.
2. **Data Analytics:** Innovations in data collection, analysis, and modeling provide policymakers with accurate and real-time information to support decision-making.
3. **Climate Modeling:** Advanced climate models enable policymakers to assess the impacts of policy choices, guiding effective mitigation and adaptation strategies.

Accelerating Positive Change:

1. **Technology and Sustainable Solutions:** Innovations in renewable energy, carbon capture, and sustainable agriculture provide practical tools for addressing

pressing challenges.

2. **Medical Breakthroughs:** Scientific advancements in healthcare, such as new treatments, diagnostics, and therapies, lead to improved public health outcomes.

3. **Efficiency Gains:** Innovations in industries ranging from manufacturing to transportation optimize processes, reduce waste, and lower carbon footprints.

Shaping Progressive Policies:

1. **Environmental Regulations:** Science-backed data on air and water quality, as well as the impacts of pollution, drive the creation and enforcement of environmental regulations.

2. **Climate Agreements:** International climate agreements, like the Paris Agreement, rely on scientific consensus to set targets and hold nations accountable.

3. **Public Health Measures:** Policymakers rely on scientific research to formulate health guidelines and responses to disease outbreaks.

Transcending Boundaries:

1. **Global Collaboration:** Scientific collaboration across borders fosters the sharing of knowledge, experiences, and best practices to address global challenges.

2. **Science Diplomacy:** Bridging international relations and scientific cooperation, science diplomacy plays a role in resolving disputes and promoting shared goals.

Addressing Ethical Concerns:

1. **Ethical Frameworks:** Ethical considerations guide the responsible use of technology, ensuring innovation aligns with societal values and principles.

2. **Equity and Inclusivity:** Science and innovation must be accessible to all, and policies should address disparities in access and benefits.

Public Awareness and Education:

1. **Science Communication:** Effective communication of scientific findings helps the public understand complex issues and empowers informed decision-making.
2. **Education Initiatives:** Science education equips future generations with the knowledge and critical thinking skills needed to engage with evolving challenges.

Challenges and Opportunities:

1. **Funding and Support:** Adequate funding is crucial for research and innovation, requiring commitment from governments, philanthropic organizations, and the private sector.
2. **Interdisciplinary Collaboration:** Tackling complex challenges often requires collaboration between diverse fields and experts.
3. **Adapting to Change:** Science and innovation need to continuously evolve to address emerging challenges, ensuring solutions remain relevant.

A Shared Responsibility: The potential for science and innovation to drive positive change and shape policy is immense, but it requires collaboration among researchers, policymakers, industry leaders, and society as a whole. By fostering a culture that values evidence-based decision-making, invests in research and development, and promotes the responsible application of innovation, we can navigate the complexities of our world and work toward a brighter future for all.

Reflecting on the insights gained from the exploration of climatology and paleoclimatology

Journeying Through Clime Chronicles: A Reflection on Climatology and Paleoclimatology

As we conclude our expedition through the realms of climatology and paleoclimatology, we stand at the crossroads of scientific discovery and our collective responsibility to the planet. The insights gained from this exploration not only deepen our understanding of Earth's intricate climate systems but also illuminate the path toward a more sustainable and resilient future.

A Tapestry of Complexity: Our voyage through climatology unveiled the intricate threads that weave together the tapestry of our planet's climate. From the delicate balance of atmospheric patterns to the complex dance of ocean currents, we discovered the interconnectedness of natural processes that shape our environment. These processes transcend borders, emphasizing the global nature of climate challenges and the need for collaborative action.

Echoes from the Past: Our journey into paleoclimatology, the study of ancient climates, offered a glimpse into the eons that came before us. Through ice cores, sediment layers, and fossil records, we traced the footprints of Earth's history. These echoes from the past served as a reminder that our planet's climate has undergone profound shifts long before our time, driving home the realization that change is a fundamental part of our planet's story.

A Call to Action: Our expedition holds a mirror to the present and beckons us to respond to the pressing challenges before us. We have witnessed the impact of human activities on the climate, from carbon emissions altering the composition of our atmosphere to rising sea levels encroaching upon coastal communities. The knowledge we have gained serves as a call to action—a call to rethink our practices, innovate our solutions, and champion policies that safeguard our planet for generations to come.

Science as Our North Star: Throughout this journey, science has been our guiding light. Rigorous research, innovative technologies, and evidence-based policymaking have illuminated our path and illuminated the way forward. Yet, science alone is not enough. It is our collective commitment, our capacity for empathy, and our willingness to act that will propel us toward a sustainable and harmonious future.

Empowering Hope and Resilience: Amidst the challenges, our exploration has also revealed stories of hope and resilience. We have witnessed communities adapting to changing climates, scientists pioneering innovative solutions, and nations uniting under the banner of global cooperation. These narratives inspire us to believe in the potential of human ingenuity and collaboration to overcome adversity.

A New Chapter Awaits: Our voyage through climatology and paleoclimatology marks just the beginning of our journey. Armed with knowledge, armed with the memory of Earth's past and the urgency of its present, we embark on a new chapter—one where each decision, each action, reverberates through the intricate web of life on this planet.

As we reflect upon the insights gathered and bid adieu to this expedition, we carry with us a profound sense of responsibility. Let our journey be a testament to our shared commitment to protect and nurture the Earth, to cherish its wonders, and to preserve the delicate balance that sustains us all.

Emphasizing the importance of responsible climate stewardship for future generations

Guardians of the Earth: Nurturing Responsible Climate Stewardship for Future Generations

In the symphony of time, we stand as mere notes, resonating across generations. As we reflect on our journey through the intricate realms of climate science, we are entrusted with a profound responsibility—the guardianship of our planet's climate for the well-being of those who will follow in our footsteps.

A Precious Inheritance: The legacy of Earth's climate is a treasure passed down from the eons gone by. Yet, it is not an immutable inheritance; it is a living, breathing system shaped by the choices we make today. What we leave behind is not just the world as it is now, but the potential for flourishing or faltering—depending on whether we choose to be responsible stewards of our planet.

Shaping Tomorrow's Landscape: The journey of responsible climate stewardship begins with understanding that our actions today mold the landscape of tomorrow. The emissions we release into the atmosphere, the resources we consume, the policies we advocate for—all of these threads weave into the fabric of our planet's future. It is our duty to ensure that this tapestry is one of vibrancy, diversity, and harmony.

Empowering Future Voices: As we tread the path of stewardship, we must amplify the voices of those who stand to inherit the consequences of our decisions. The young, the yet-to-be-born—they too have a stake in the climate story. Let us listen to their calls for action, for justice, and for a world that respects the balance of nature.

Choosing Sustainability over Convenience: In our modern age, convenience often comes at a cost to the environment. Yet, the true cost is borne by the generations to come. Responsible climate stewardship challenges us to prioritize sustainability over short-term convenience. It beckons us to embrace alternative energy sources, reduce waste, and make choices that safeguard the fragile equilibrium of our ecosystems.

Innovating Solutions, Fostering Hope: Innovation is our ally in this endeavor. As we delve into renewable energies, carbon capture technologies, and climate-resilient infrastructure, we kindle a beacon of hope. Each innovation carries the promise of a future unburdened by the weight of unsustainable practices.

Harmony, Not Discord: Our planet is a delicate symphony of interconnected elements. As responsible stewards, we must strive for harmony rather than discord. Our actions should be a chorus that sings in tune with nature's rhythms, rather than a cacophony that disrupts the melody.

A Collective Symphony: Responsibility knows no borders, and stewardship is a shared endeavor. Every nation, every community, every individual contributes to the composition of this collective symphony of climate stewardship. It is in our shared commitment that we find the strength to overcome challenges and shape a harmonious future.

The Promise We Hold: We stand at a crossroads—a junction of choices that will ripple across time. The path of responsible climate stewardship is not just about preserving the environment; it's about honoring the promise we hold to the future. It's about leaving a legacy of compassion, foresight, and resilience.

As we write the next chapters of Earth's story, let us pen a narrative of responsible climate stewardship. Let us pass on a planet that is thriving, abundant, and enriched by our collective commitment to safeguarding its beauty and vitality. The echoes of our actions will resound through the ages, and in those echoes, future generations will find the heartbeat of a world nurtured with care and love.

Encouraging readers to engage in ongoing learning, awareness, and action to safeguard Earth's climate

Igniting Change: Embrace the Journey of Lifelong Climate Advocacy

As we draw the curtain on our exploration of Earth's climate, let us not bid farewell to the cause that has ignited our curiosity and stirred our sense of responsibility. Instead, let this be a new beginning—a call to arms, a rallying cry, and an unwavering commitment to safeguarding our planet's climate for generations to come.

A Call to Lifelong Learning: Our journey through climate science has illuminated the depths of knowledge that lie before us. Yet, this knowledge is ever-evolving, and our quest for understanding should never cease. Let us embrace the joy of learning, whether through scientific literature, documentaries, or engaging with experts who can guide us on this path of enlightenment.

Fueling the Fire of Awareness: With knowledge comes awareness, and with awareness comes the power to inspire change. Let us be the catalysts of conversations that echo the urgency of climate action. Share your insights, engage with friends and family, and foster a community of climate-conscious individuals united by a common cause.

Small Steps, Big Impact: Change need not be monumental to be meaningful. Every action—whether reducing single-use plastics, conserving energy, or supporting sustainable products—contributes to the greater good. Our collective small steps can lead to a giant leap toward a sustainable future.

Empowering Advocacy: Our journey through climate science equips us with a powerful tool—knowledge. Armed with this tool, we have the capacity to influence policies, demand sustainable practices, and hold leaders accountable. Let us not shy away from advocacy, for our voices are essential in driving systemic change.

Embrace Innovation and Solutions: Innovation thrives on collaboration, and the climate cause is no exception. Explore renewable energy options, support startups focused on climate solutions, and engage with organizations dedicated to reversing climate change. By investing in innovation, we invest in hope.

Taking the First Step: The journey of climate advocacy begins with a single step—an acknowledgment of the urgency, a recognition of the responsibility, and a commitment to action. Let this be your moment to step onto this path, armed with the knowledge that your efforts contribute to a healthier planet.

Never Underestimate Your Impact: Every individual has the potential to be a force for change. Whether you inspire a friend to make sustainable choices or join a local climate action group, remember that your actions create ripples that can shape a tide of transformation.

A Pledge to Future Generations: As you stand at the intersection of knowledge and action, envision the world you wish to leave for those who will follow. Pledge to be a guardian of Earth's climate, to steward its beauty, and to ensure that future generations inherit a planet that is vibrant, abundant, and harmonious.

Let Your Journey Continue: Our exploration of Earth's climate is not a destination; it is a journey that stretches ahead, beckoning us to discover, learn, and act. As you close this chapter, remember that the pages of this journey are endless. Let your curiosity be your guide, your knowledge your compass, and your actions a testament to the indomitable spirit of humanity in the face of challenge.

The journey continues, the cause remains vital, and the planet's fate rests in our hands. Embrace the journey of lifelong climate advocacy, for in this journey lies the hope of a thriving and sustainable world for all.

Acknowledging the ever-evolving nature of climatology

Evolving Horizons: Embracing the Dynamic Nature of Climatology

In the vast expanse of time, the study of climatology stands as a testament to the ever-changing nature of our planet. As we reflect on our journey through this intricate science, let us pause to acknowledge the fluidity and adaptability that define climatology.

A Field in Flux: Climatology, like the climate it examines, is in constant motion. It evolves as our understanding deepens, as new technologies emerge, and as the challenges of our era unfold. This very nature is a reminder that the pursuit of knowledge is a journey, not a destination.

Adapting to the Unpredictable: The climate is a symphony of variables that orchestrate intricate patterns. Our ability to predict, adapt, and respond to these patterns relies on the evolving landscape of climatology. The challenges of today may be vastly different from those of tomorrow, and our field must be prepared to navigate uncharted waters.

Championing Innovation: In the face of uncertainty, innovation becomes our ally. We develop new methodologies, harness cutting-edge technologies, and cultivate interdisciplinary collaborations that enrich our understanding of climate dynamics. These innovations hold the key to unlocking solutions for a sustainable future.

A Canvas of Data: As we journey through the chapters

of climatology, we accumulate a rich tapestry of data. This data paints a vivid picture of Earth's climate story, revealing trends, shifts, and anomalies that shape our world. Yet, as we acknowledge the value of this data, we also recognize its limitations and the potential for continuous refinement.

Collaborative Endeavors: Climatology thrives on collaboration —a collective effort that transcends boundaries. Scientists, policymakers, communities, and individuals come together to build a mosaic of insights that informs our responses to climate challenges. This shared pursuit is a testament to the unity needed to address global issues.

A Humble Role in Nature's Play: Our exploration of climatology is a mere glimpse into the grand theater of Earth's climate systems. As we observe, analyze, and interpret, we remember that our role is both profound and humble—an attempt to decipher the complex interactions that have shaped our planet for millennia.

An Invitation to Curiosity: The evolving nature of climatology extends an invitation to curiosity—a call to continuously question, learn, and adapt. With every discovery, we realize that there is always more to learn, more to uncover, and more to contribute to the legacy of knowledge.

Cherishing Progress and Potential: As we close this chapter of exploration, let us cherish the progress made and the potential that lies ahead. Let us recognize that our journey through climatology is ongoing, and as we traverse the uncharted territories of the future, we carry with us the lessons of the past.

A Final Thought: Climatology, like the very climate it studies, is a dynamic force—an intricate dance of observation, analysis, and adaptation. It is a journey that extends beyond our lifetimes, a pursuit that echoes the resounding truth that the only constant is change itself.

Inspiring readers to carry the knowledge and lessons of climatology into their lives and communities

Bridging Knowledge to Action: Empowering Change Through Climatology

As we conclude our exploration of climatology, we are not bidding farewell to a subject; we are embarking on a journey of transformation—an opportunity to integrate the knowledge we've gained into our lives and communities. The lessons of climatology are not meant to remain confined to the pages of textbooks; they are meant to be catalysts for change, engines of awareness, and sparks of inspiration that ignite positive action.

From Theory to Reality: The insights we've gathered are more than intellectual pursuits; they are seeds waiting to be sown in the fertile soil of our lives. Let us bridge the gap between theory and reality, transforming knowledge into action that resonates in our choices, behaviors, and contributions.

Championing Awareness: The power of awareness cannot be underestimated. Let the knowledge of climatology be your lantern, guiding conversations with friends, family, and colleagues. Share the urgency, the solutions, and the need for collective action. Through these conversations, you can foster a ripple effect of consciousness.

Sustainable Living, Everyday Impact: Simple choices can yield profound results. Embrace sustainable practices in your daily life—reduce energy consumption, minimize waste, support local initiatives, and choose products with a lower environmental footprint. Your actions, however small, accumulate to shape a more sustainable world.

Community Engagement: A community united in purpose is a force to be reckoned with. Organize workshops, seminars, and local events to raise awareness about climate issues. By engaging with your community, you amplify your impact and inspire others to take action.

Empowering Youth: The young minds of today will shape tomorrow's world. Encourage young individuals to explore climatology, engage in discussions, and develop innovative solutions. Mentorship and education can cultivate a generation of climate advocates.

Advocacy and Policy: Knowledge is the foundation of effective advocacy. Use your understanding of climatology to advocate for policies that prioritize environmental protection, renewable energy, and climate resilience. Urge your representatives to prioritize sustainable measures.

Adaptation and Resilience: Incorporate climate-conscious thinking into your future plans. Advocate for resilient infrastructure, urban planning that considers climate impacts, and sustainable agricultural practices. This forward-thinking approach safeguards communities against climate shocks.

Inspire Others to Act: Your passion is contagious. Share your journey of learning, reflection, and action with others. By being an example of change, you inspire those around you to embrace their own roles as climate stewards.

Lifelong Learning: Climatology is a dynamic field, always evolving. Commit to a lifelong journey of learning—stay updated on the latest research, technologies, and breakthroughs. This continual engagement ensures that your actions remain well-informed and effective.

A Collective Legacy: As we embark on this new phase of our journey, let us remember that the legacy of climatology is not confined to textbooks—it's etched in the choices we make, the communities we influence, and the planet we protect. Carry the lessons of climatology into your life, infuse them into your interactions, and nurture the seeds of change that you've sown.

Climatology isn't a subject to master and leave behind; it's an ongoing commitment to understanding, respecting, and safeguarding our planet. Let the knowledge you've gained be the wind beneath your wings as you soar toward a future where every action, every decision, and every heartbeat resonates with the rhythm of responsible climate stewardship.

A call to continue exploring, learning, and advocating for a resilient and sustainable planet

Answering the Call: Uniting for a Resilient and Sustainable Planet

Our journey through the realms of climate science has been a voyage of discovery, enlightenment, and empowerment. Yet, as we draw this chapter to a close, let us recognize that our expedition is far from over. It is a call to continue exploring, learning, and advocating for a planet that thrives, a call that resonates with urgency and hope.

A Never-Ending Exploration: The universe of climate science is vast, and our journey has merely scratched the surface. The intricacies of Earth's climate are waiting to be uncovered, the solutions to our challenges are yet to be fully realized, and the stories of resilience and innovation continue to unfold. Let our curiosity be insatiable, driving us to delve deeper into this ever-evolving landscape.

In the Company of Lifelong Learners: Our journey has been shared with pioneers, researchers, and thinkers who have dedicated their lives to understanding the nuances of our climate. Let us remember that learning is not a solitary endeavor. It is a chorus of voices, a symphony of collaboration, and an invitation to stand shoulder to shoulder with fellow lifelong learners.

A Beacon of Advocacy: Our knowledge is a beacon that guides us toward a brighter future. Armed with this knowledge, let us advocate tirelessly for policies that promote sustainability, for innovations that mitigate our impact on the planet, and for equitable solutions that benefit all corners of society. Our voices, collectively raised, can drive transformative change.

Champions of Resilience: The lessons of our journey inspire us to be champions of resilience—individuals who recognize that our actions today shape the world of tomorrow. Let us fortify communities against the impacts of climate change, nurture ecosystems that sustain life, and embrace adaptability as a cornerstone of our existence.

A Legacy for Generations: Our commitment to a resilient and sustainable planet is a legacy that we leave for generations yet to come. It is a testament to our shared vision of a world where the delicate balance of nature is preserved, where ecosystems flourish, and where the human spirit thrives in harmony with the environment.

A Pledge to Action: As we embark on the next phase of our journey, let us make a pledge to ourselves, our communities, and our planet. Let us commit to continually expanding our knowledge, fostering awareness, and advocating for change. Let us embody the spirit of responsibility, stewardship, and unwavering dedication.

A Future of Possibility: The path before us is both a challenge and an opportunity. It is a canvas upon which we can paint a vision of a world united by the common goal of a resilient and sustainable planet. As we take each step, let us remember that our journey is guided by the aspirations of generations past, the urgency of the present, and the promise of a better future.

Answering the Call Together: Our journey through climate science is not an individual endeavor; it is a collective call to action. It is a call to scientists, policymakers, educators, parents, and individuals of all walks of life. Let us heed this call and stand united in our pursuit of a world that thrives, endures, and embodies the very essence of resilience and sustainability.

With hearts ablaze and minds open, let us continue exploring, learning, and advocating—for our planet, for ourselves, and for the generations yet to be. Our journey is far from over; it is a marathon fueled by determination, guided by knowledge, and steered by the unwavering belief that together, we can sculpt a future that embraces the boundless potential of a resilient and sustainable planet.